INTEGRALS CONTAINING $\sqrt{X^2 - A^2}$

$$\int \sqrt{x^2 - a^2}\,dx = \frac{1}{2}x\sqrt{x^2 - a^2} - \frac{1}{2}a^2 \log\left|x + \sqrt{x^2 - a^2}\right|$$

$$\int \frac{\sqrt{x^2 - a^2}}{x}\,dx = \sqrt{x^2 - a^2} - a\sec^{-1}\frac{x}{a}$$

$$\int \frac{dx}{\sqrt{x^2 - a^2}} = \log\left|x + \sqrt{x^2 - a^2}\right| = \cosh^{-1}\frac{x}{a}$$

$$\int \frac{dx}{x\sqrt{x^2 - a^2}} = \frac{1}{a}\sec^{-1}\frac{x}{a}$$

INTEGRALS CONTAINING $\sqrt{A^2 + X^2}$

$$\int \sqrt{a^2 + x^2}\,dx = \frac{1}{2}x\sqrt{a^2 + x^2} + \frac{1}{2}a^2 \log\left|x + \sqrt{a^2 + x^2}\right|$$

$$\int \frac{dx}{\sqrt{a^2 + x^2}} = \log\left|x + \sqrt{a^2 + x^2}\right|$$

$$\int \frac{dx}{x\sqrt{a^2 + x^2}} = \frac{1}{a}\log\left|\frac{\sqrt{a^2 + x^2} - a}{x}\right|$$

$$\int \frac{dx}{(a^2 + x^2)^{3/2}} = \frac{x}{a\sqrt{a^2 + x^2}}$$

OTHER INTEGRALS

$$\int xe^x\,dx = e^x(x - 1)$$

$$\int x^2e^x\,dx = e^x(x^2 - 2x + 2)$$

$$\int \log x\,dx = x\log x - x$$

$$\int x\log x\,dx = \frac{1}{2}x^2\log x - \frac{1}{4}x^2$$

$$\int e^x \sin x\,dx = \frac{1}{2}e^x(\sin x - \cos x)$$

$$\int e^x \cos x\,dx = \frac{1}{2}e^x(\sin x + \cos x)$$

$$\int x\sin x\,dx = \sin x - x\cos x$$

$$\int x\cos x\,dx = \cos x + x\sin x$$

$$\int \sin^{-1} x\,dx = x\sin^{-1} x + \sqrt{1 - x^2}$$

$$\int \cos^{-1} x\,dx = x\cos^{-1} x - \sqrt{1 - x^2}$$

$$\int \tan^{-1} x\,dx = x\tan^{-1} x - \frac{1}{2}\log(1 + x^2)$$

UNDERSTANDING CALCULUS

IEEE Press
445 Hoes Lane, P.O. Box 1331
Piscataway, NJ 08855-1331

IEEE Education Society, *Sponsor*
E-S Liaison to IEEE Press, Robert J. Herrick

Cover Design: Caryl Silvers, *Silvers Design*

Technical Reviewers

Susan A. R. Garrod, *Purdue University, IN*
Glen E. Harland Jr., *Purdue University, IN*
Robert J. Herrick, *Purdue University, IN*
Timothy N. Trick, *University of Illinois at Urbana-Champaign*

Books of Related Interest from the IEEE Press

THE CALCULUS TUTORING BOOK
C. Ash and R. Ash
1986 Softcover 544 pp IEEE Order No. PP1776 ISBN 0-7803-1044-6

ONES AND ZEROS: Understanding Boolean Algebra, Digital Circuits, and the Logic of Sets
John R. Gregg
1998 Softcover 296 pp IEEE Order No. PP5388 ISBN 0-7803-3426-4

THE PROBABILITY TUTORING BOOK: Intuitive Essentials for Engineers and Scientists and Everyone Else!
Carol Ash
1993 Softcover 480 pp IEEE Order No. PP2881 ISBN 0-7803-1051-9

UNDERSTANDING CALCULUS

A User's Guide

H. S. Bear
Professor Emeritus
Department of Mathematics
University of Hawaii

IEEE Education Society, *Sponsor*

IEEE Press Understanding Science & Technology Series

The Institute of Electrical and Electronics Engineers, Inc., New York

This book and other books may be purchased at a discount
from the publisher when ordered in bulk quantities. Contact:

IEEE Press Marketing
Attn: Special Sales
445 Hoes Lane
P.O. Box 1331
Piscataway, NJ 08855-1331
Fax: +1 732 981 9334

For more information about IEEE Press products, visit the
IEEE Online Catalog & Store: http://www.ieee.org/ieeestore.

Printed in the United States of America.

10 9 8 7 6 5 4 3 2 1

ISBN 0-7803-6018-4
IEEE Order No. PP5867

Library of Congress Cataloging-in-Publication Data

Bear, H. S. (Herbert Stanley)
Understanding calculus : a user's guide / H.S. Bear.
p. cm. -- (IEEE Press Understanding Science & Technology Series)
Includes index.
ISBN 0-7803-6018-4
1. Calculus. I. Title. II. Series.

QA303 .B38 2001
515--dc21

00-033522

Contents

Author's Message to the Reader

Calculus is simple, and it's fun!

Well, OK, that's a slight exaggeration, but it's not as far off as you think. Calculus does have a reputation of being a difficult subject, but it's a bad rap. What is true is that calculus textbooks are impossibly obscure, and consequently calculus courses are unnecessarily difficult.

Most traditional calculus textbooks exceed 1,000 pages, weigh nearly 10 pounds, cost $100 or more, and contain more material than a normal person could digest in four years. These textbooks are written for the instructor rather than the student, because it is the instructor who chooses the text. Thus, all the texts include every possible topic any instructor might want to cover, making them impossible for the average reader to navigate.

The typical calculus instructor is a devout mathematician and is deeply concerned with mathematical doctrine. Definitions, theorems, and proofs are the essence of advanced mathematics, and these are what the instructor wants in a calculus text. The difficulty is that abstract mathematics makes no sense until one has a command of many specific examples. The beginning student of calculus is simply not prepared to deal with the kind of formalism that most texts present.

The purpose of this book is to show you how calculus works, and to provide enough concrete examples and exercises to enable you to work with calculus. The material is illustrated by simple physical or geometric examples. Formal theorems are replaced by helpful descriptions, and proofs are replaced by explanations. These explanations can easily be turned into proofs by anyone who is so inclined, but most readers will be content with examples on their first attempt. Nothing in this book will ever have to be unlearned. It isn't formal, but it is correct.

Although calculus instructors love to lecture on theorems and proofs, they do bow to reality and stick largely to computational problems on examinations. Most of the problems that occur on any calculus test are found in this book. If you can solve these problems, you have learned what a calculus student is expected to know.

So, if you've taken calculus in the past and want to review it, or if you need help right now in your calculus course in deciding what is important, or if you're looking for a different approach to help understand a topic, this book is what you need.

ANNOTATED TABLE OF CONTENTS

Chapter 1. Lines

This chapter introduces the Cartesian coordinate system for the plane and explains the idea of the graph of an equation in x and y as a curve in the plane. The simplest curves are straight lines, which are the graphs of linear equations $Ax + By + C = 0$. We show how to graph a given equation and how to write the equation of a line described geometrically.

Chapter 2. Parabolas, Ellipses, Hyperbolas

These curves, the so-called *conic sections*, are the graphs of second-degree equations in x and y. Every curve $y = Ax^2 + Bx + C = 0$ is a parabola, every curve $x^2 + y^2 + Ax + By + C = 0$ is a circle, and so on. The emphasis here is on being able to sketch a given curve quickly, for these curves play a large part in how we visualize the facts of calculus.

Chapter 3. Differentiation

The derivative is introduced as the kind of computation used to calculate speed, given a formula for distance in terms of time, or to calculate the slope of a curve $y = f(x)$. The idea of a limiting value is treated intuitively, and several calculations involving limits are made before an informal definition of limit is presented. The exercises involve calculating derivatives for simple functions with applications to tangent lines, speeds, and other rates of change.

Chapter 4. Differentiation Formulas

The notation Δx for a small change in x and Δy for the corresponding change in y is used in the arguments to justify the rules for differentiating sums, products, and so on, and to presage the $\frac{dy}{dx}$ notation. The product rule and informal induction yield the formula $\frac{d}{dx}x^n = nx^{n-1}$ for positive and negative integers n. The case $n = \frac{1}{2}$ is also covered to foreshadow the rule for noninteger exponents. The limit theorems for sum, product, and so on are taken as obvious properties of the arithemetic operations.

Chapter 5. The Chain Rule

Examples of composite functions are given, using both functional notation and dependent variable notation. For example, $g(f(x)) = (2x^2 + 1)^3$, with $g(x) = x^3$ and $f(x) = 2x^2 + 1$, or $y = z^3$ with $z = 2x^2 + 1$. The dependent variable notation provides the clearest explanation of the chain rule, writing $\frac{\Delta y}{\Delta x} = \frac{\Delta y}{\Delta z} \cdot \frac{\Delta z}{\Delta x}$ and taking limits. Related rate problems are introduced where, for example, a relationship between distances s and x is used to find $\frac{ds}{dt}$ from a given $\frac{dx}{dt}$.

Chapter 6. Trigonometric Functions

The functions $\cos x$ and $\sin x$ are defined to be the coordinates of the point x units around the unit circle from $(1, 0)$. The distance x is also the radian measure of the angle between the positive x-axis and the radius to the point $(\cos x, \sin x)$. A table of degree-radian equivalencies and a list of the basic trigonometric identities are given to be memorized. The derivatives of $\cos x$, $\sin x$, $\tan x$, and $\sec x$ are calculated and illustrated with many chain rule examples and rate problems.

Chapter 7. Exponential Functions and Logarithms

The general shape of $y = a^x$ for $a > 1$ is explained, and it is shown that $\frac{d}{dx} a^x = k_a a^x$ for some constant k_a depending on a. We define e to be the number such that $k_e = 1$, so $\frac{d}{dx} e^x = e^x$, and it is shown that $e \doteq 2.718$. The function $\log x$ is the inverse of e^x, and the derivative $\frac{d}{dx} \log x = \frac{1}{x}$ is calculated from the inverse relationship. Many chain rule exercises involving e^x and $\log x$ are given, as well as some problems involving the exponential growth of bacteria colonies and compound interest.

Chapter 8. Inverse Functions

General inverse functions are defined with e^x and $\log x$ as a convenient example. The nth root of x, $x^{\frac{1}{n}}$, is the inverse of x^n for $x > 0$. The derivative of $y = x^{\frac{1}{n}}$ is obtained by differentiating both sides of $y^n = x$, and the chain rule then extends the differentiation formula to fractional exponents. The functions $\sin^{-1} x$, $\cos^{-1} x$, $\tan^{-1} x$ are defined, and their derivatives are calculated. For example, if $y = \sin^{-1} x$, we differentiate both sides of $\sin y = x$ and then use the appropriate right triangle to show that $\cos(\sin^{-1} x) = \sqrt{1 - x^2}$.

Chapter 9. Derivatives and Graphs

The first and second derivatives are used to determine the shape of a curve. For example, the function is increasing if $f'(x) > 0$, and concave up if $f''(x) > 0$. Local maxima and minima can only occur where $f'(x) = 0$. Implicit differentiation is used to find the slope and convexity of a function defined by an equation. Applications are given to curve sketching and stated max/min problems.

Chapter 10. Following the Tangent Line

L'Hospital's Rule for limits of the form $\frac{0}{0}$ is derived quite simply, using the definition of $f'(x_0)$ and $g'(x_0)$ to show that if $f(x_0) = g(x_0) = 0$, then $\frac{f(x)}{g(x)} \to \frac{f'(x_0)}{g'(x_0)}$ as $x \to x_0$. Newton's method of solving an equation $f(x) = 0$ consists of starting with an approximate solution of x_1 and following the tangent line back to the x-axis to get the better approximation $x_2 = x_1 - \frac{f(x_1)}{f'(x_1)}$.

Chapter 11. The Indefinite Integral

Indefinite integration is the operation inverse to differentiation, so $F(x)$ is the integral or antiderivative of $f(x)$ if $F'(x) = f(x)$. We start with the problem of turning the gravitational equation $\frac{d^2 s}{dt^2} = g$ into a formula for s in terms of t. All the standard differentiation formulas are listed alongside the corresponding integration formulas. Some simple differential equations with initial conditions are solved. The u-substitution is used to explain the backwards use of the chain rule in integration problems.

Chapter 12. The Definite Integral

The area between the curve $y = f(x)$ and the x-axis, for $a \le x \le b$, is defined to be the limit of the Riemann sums $\sum f(c_i)(x_i - x_{i-1})$ where the limit is taken as $\max(x_i - x_{i-1}) \to 0$. This limit is the definite integral, denoted $\int_a^b f(x)dx$. The Mean Value Theorem shows that $\int_a^b f(x)dx = F(b) - F(a)$ for any antiderivative $F(x)$ of $f(x)$. Areas between curves are calculated using the mnemonic that the area is the sum, indicated by the integral sign, of vertical "rectangles" of height $f(x)$ and width dx. For curves $x = f(y)$, one uses horizontal "rectangles" to get the area in the form $\int_a^b f(y)dy$.

Chapter 13. Work, Volume, Force

Work, volume, and force are calculated as integrals that represent limits of sums of increments. If a force at point x is $f(x)$, then an increment of work done by this force is $f(x)dx$, and the total work is the sum (integral) of these increments. Similarly, if a solid has cross-sectional area $A(y)$ at height y, then an increment of volume is $A(y)dy$, and total volume is the integral $\int A(y)dy$. Many standard work/volume/force problems are treated.

Chapter 14. Parametric Equations

The x and y coordinates of a moving particle are frequently given as functions of t with equations $y = f(t)$, $x = g(t)$. Two such equations are called the *parametric equations of the path.* Many curves are most conveniently described by parametric equations, where the parameter could be time, or perhaps some variable of geometric importance to the curve. The slope of a parametrically given curve is $\frac{dy}{dx} = \frac{\frac{dy}{dt}}{\frac{dx}{dt}}$, and the concavity $\left(\frac{d^2y}{dx^2}\right)$ is also calculated in terms of t. Cauchy's Mean Value Theorem is the ordinary Mean Value Theorem applied to parametric curves.

Chapter 15. Change of Variable

When a u-substitution is used to change a definite integral in x directly into a definite integral in u, the process is called *a change of variable.* If the interval of integration for the original integral is $a \le x \le b$, and $u = c$ when $x = a$ and $u = d$ when $x = b$, then the u-integral will have lower limit c and upper limit d (even if $d < c$). The new integral formulas $\int (a^2 + x^2)^{-1} dx = \frac{1}{a} \tan^{-1} \frac{x}{a}$ and $\int (a^2 - x^2)^{-\frac{1}{2}} dx = \sin^{-1} \frac{x}{a}$ are introduced. There are lots of examples of the technique.

Chapter 16. Integrating Rational Functions

A rational function $\frac{P(x)}{Q(x)}$, where $P(x)$ and $Q(x)$ are polynomials, can always be integrated if $Q(x)$ can be completely factored. We consider the most common and most useful cases; namely, $Q(x)$ is linear or quadratic, and the degree of $P(x)$ is less that the degree of $Q(x)$.

Chapter 17. Integration by Parts

The integration by parts technique uses the differentiation formula $\frac{d}{dx}(uv) = u\frac{dv}{dx} + v\frac{du}{dx}$ in the integrated form $uv = \int udv + \int vdu$. This equation can be used to write one integral in terms of another: $\int udv = uv - \int vdu$. We cover the principal examples where $\int vdu$ is simpler than $\int udv$, and the technique works. Integration by parts is used to integrate $\log x$ and the inverse trigonometric functions.

Chapter 18. Trigonometric Integrals

Trigonometric functions are a basic part of scientific language, it is essential to be able to integrate formulas involving these functions. Trigonometric integrals also arise when substitutions are made to integrate expressions involving radicals. The basic trigonometric identities are listed for convenient memorization—and, yes, memorization really is necessary here. The basic techniques are covered for integrals with sines and cosines, and integrals with secants and tangents.

Chapter 19. Trigonometric Substitution

Integrands that contain the radical expressions $\sqrt{a^2 - x^2}$, $\sqrt{a^2 + x^2}$, $\sqrt{x^2 - a^2}$ can frequently be integrated after a trigonometric substitution. The substitution $x = a \sin \theta$, for example, turns $\sqrt{a^2 - x^2}$ into $a \cos \theta$, using the identity $1 - \sin^2 \theta = \cos^2 \theta$. After integration, the appropriate right triangle is used to convert trigonometric functions of θ back to formulas in x. Both definite and indefinite integrals are treated.

Chapter 20. Numerical Integration

There are several methods to get a numerical approximation to $\int_a^b f(x)dx$ using different ways to choose the c_i in the Riemann sums $\sum f(c_i)(x_i - x_{i-1})$. These methods emphasize the fact that the integral is the limit of Riemann sums, but all such methods are all vastly less efficient than Simpson's Rule, which is the only approximation method we consider. Simpson's Rule approximates short segments of the curve by parabolic arcs, finds the exact areas under the parabolic arcs, and adds them up in one simple formula. In some cases, Simpson's Rule is less tedious than standard methods for getting the exact answer.

Chapter 21. Limits at ∞; Sequences

We consider the limits of the form $\lim_{x \to \infty} f(x)$, where $f(x)$ is defined on some interval (a, ∞), and limits $\lim_{n \to \infty} x_n$, where $\{x_n\}$ is a sequence. For rational functions $\frac{P(x)}{Q(x)}$ the limit is simply $\lim_{n \to \infty} \frac{a_n x^n}{b_m x^m}$, where $a_n x^n$ and $b_m x^m$ are the highest order terms of $P(x)$, $Q(x)$, respectively. L'Hospital's Rule for the indeterminate forms $\frac{0}{0}$ and $\frac{\infty}{\infty}$ is used to show that $\frac{\log x}{x^p} \longrightarrow 0$ if $p > 0$, and $\frac{x^p}{a^x} \longrightarrow 0$ if $a > 1$. The four sequences $\{\log n\}$, $\{n^p\}$, $\{a^n\}$, $\{n!\}$ represent distinct orders of magnitude as $n \longrightarrow \infty$ so that $\frac{\log n}{n^p} \longrightarrow 0$, $\frac{n^p}{a^n} \longrightarrow 0$, $\frac{a^n}{n!} \longrightarrow 0$. These limits are critical in our later treatment of power series.

Chapter 22. Improper Integrals

Initially, the definite integral $\int_a^b f(x)dx$ is defined only for bounded intervals $[a, b]$ and bounded functions $f(x)$. Now we extend the definition to cover functions that are unbounded on $[a, b]$, and integrals over unbounded intervals like $(-\infty, b]$ or $[a, \infty)$. These new (improper) integrals are naturally defined as limits of the original (proper) integrals. Particular emphasis is placed on the integrals of the form $\int_a^\infty f(x)dx$, with $f(x) > 0$, since these integrals occur most often and have important applications to infinite series.

Chapter 23. Series

The sums of infinitely many terms, indicated $a_1 + a_2 + \cdots + a_n + \cdots$, is defined to be the limit of the sums of the first n terms. This definition depends on the order in which the terms are listed, and changing that order can affect the value of the sum. For a series to have the standard properties of finite sums—for instance, the sum is not affected by the order of the terms—the series must converge absolutely; that is, the series of absolute values must converge. The geometric series, $1 + x + x^2 + \cdots + x^n + \cdots$, which equals $\frac{1}{1-x}$ if $|x| < 1$, is extremely important since the convergence of a power series is usually determined by comparison with a geometric series.

Chapter 24. Power Series

Series of the form $\sum a_n(x - x_0)^n$, with x_0 possibly equal zero, are called *power series*. Such a series either converges for all x or converges on some open interval around x_0. All the basic functions of calculus can be represented by power series on some open interval. Power series can be differentiated or integrated term-by-term without changing the interval of convergence. The geometric series for $\frac{1}{1-x}$ yields series for $\frac{1}{1+x^2}$, $\frac{1}{1+x}$, $\frac{1}{1-3x}$, and so on, by simple substitutions, and these series in turn can be differentiated or integrated to get series for $\tan^{-1} x$, $\log(1 + x)$, $(1 - 3x)^{-2}$, and many others.

Chapter 25. Taylor Polynomials

The nth Taylor polynomial for $f(x)$ at x_0 is the nth degree polynomial whose derivatives at x_0 agree with those of $f(x)$ up to $f^{(n)}(x_0)$. How close this polynomial, $P_n(x)$, is to $f(x)$ on an interval depends on how big $f^{(n+1)}(x)$ is on the interval, and how close x is to x_0. There is a formula to estimate the difference $|f(x) - P_n(x)|$. Many functions do not have an elementary antiderivative and so cannot be integrated directly. We can estimate such integrals by integrating the approximations $P_n(x)$. The Taylor polynomials for $f(x)$ are the partial sums of the power series for $f(x)$.

Chapter 26. Taylor Series

If $f(x)$ is given by a series $\sum a_n x^n$, then the coefficients a_n are given by $a_n = \frac{1}{n!} f^{(n)}(0)$. This formula lets us calculate the series for a given function, and this uniquely determined series is the Taylor series for $f(x)$. The partial sums of the Taylor series are the Taylor polynomials $P_n(x)$, and we use the estimates for $|f(x) - P_n(x)|$ from the last chapter to show that $P_n \longrightarrow f(x)$. The series for e^x, $\sin x$, $\cos x$, and $(1 + x)^{\frac{1}{2}}$ are calculated and are shown to converge to the right function. The series for $(1 + x)^{\frac{1}{2}}$ is an example of the extended binomial expansion.

Chapter 27. Separable Differential Equations

A *differential equation* is an equation involving two variables and the derivative of one with respect to the other. Here we study first-order equations, which means only the first derivative occurs, and equations in which the variables can be separated: $g(y)dy = f(x)dx$. The solution of a first-order differential equation is a one-parameter family of curves, so an initial condition such as $y(x_0) = y_0$ is necessary to determine a specific solution. Examples are given to illustrate exponential growth/decay, Newton's law of cooling, and falling bodies with air resistance.

Chapter 28. First-Order Linear Equations

Linear differential equations are those in which y and its derivatives occur only to the first power and there are no cross-product terms like $y\frac{dy}{dx}$. A linear first-order equation therefore looks like this: $\frac{dy}{dx} + p(x)y = q(x)$. If $P(x) = \int p(x)dx$ and both sides of the equation are multiplied by $P(x)$, then both sides can be integrated to obtain a general formula for all solutions. If $p(x)$ is constant, the method of undetermined coefficients is the easiest approach for suitable functions $q(x)$; for example, polynomials. This approach is applied to cooling and falling body problems.

Chapter 29. Homogeneous Second-Order Linear Equations

Here we treat second-order equations of the form $\frac{d^2y}{dt^2} + a\frac{dy}{dt} + by = 0$, where a and b are constant. The independent variable is now t since many of these problems involve a time-dependent variable y. The general solution, which is a linear combination of real or complex exponentials, can be determined simply by solving a quadratic equation. An initial condition now requires specifying a value for both y and $\frac{dy}{dt}$ at some t_0. If the equation has the form $\frac{d^2y}{dt^2} + w^2y = 0$, then the motion is simple harmonic and $y = K\sin(wt + \alpha)$.

Chapter 30. Nonhomogeneous Second-Order Equations

This chapter continues the study of linear second-order equations, and now the right-hand side is a function $q(t)$. Since we know all the solutions of the reduced equation, we need only find one solution of the nonhomogeneous equation. Substituting $y = y_0 v$, where y_0 is a solution of the reduced equation, leads to an equation in v which we can solve with a couple of (possibly unattractive) integrations. As was the case with first-order equations, the method of undetermined coefficients is generally more efficient. We specify the types of functions $q(t)$ for which the method works.

H. S. Bear
University of Hawaii

Acknowledgments

The author would like to acknowledge the excellent typing and proofreading of Susan Hasegawa and Pat Goldstein.

H. S. Bear
University of Hawaii

1

Lines

One of the very pleasant things about calculus is the fact that we can draw a picture of nearly everything we do. The graph of a function or an equation gives us something concrete to look at and hang our analytic ideas on. The clever device that allows us to geometrize our ideas is of course the Cartesian coordinate system, named after the French mathematician René Descartes. The Cartesian coordinate system not only gives us a pictorial representation of the ideas, but it allows us to use algebraic methods on geometric problems and geometric methods on algebraic problems.

The coordinate system consists of two perpendicular lines, the horizontal one called the ***x*-axis** and the vertical one called the ***y*-axis**. Each point in the plane is identified by a pair of numbers (x, y), where x gives the distance to the right or left of the y-axis and y gives the distance up or down from the x-axis. If $x > 0$, then the point is x units to the right of the y-axis, and if $x < 0$, then the point is $|x|$ units to the left of the y-axis. ($|x|$ denotes the magnitude of the number x, so $|x| = x$ if $x \geq 0$ and $|x| = -x$ if $x < 0$. $|x|$ is called the **absolute value** of x.) Similarly, the point (x, y) is above the x-axis y units if $y > 0$, and below the x-axis if $y < 0$. The numbers x and y are the **coordinates** of the point (x, y). The coordinate axes divide the plane into four **quadrants** (Figure 1.1).

The graph of an equation in x and y is the set of all points whose coordinates satisfy the equation. The graph of an equation is usually a curve in the plane, and here we look at the simplest curves, straight lines.

If (x_1, y_1) and (x_2, y_2) are any two points on a line, then the quantity $m = (y_2 - y_1)/(x_2 - x_1)$ is called the **slope** of the line. You get the same slope, m, no matter what two points you choose. Hence, if (x_1, y_1) is any point on a given line with slope m, and (x, y) is any other point, then

$$\frac{y - y_1}{x - x_1} = m,$$

or

$$y - y_1 = m(x - x_1). \tag{1.1}$$

Equation (1.1) therefore characterizes the line through (x_1, y_1) with slope m. If the line with

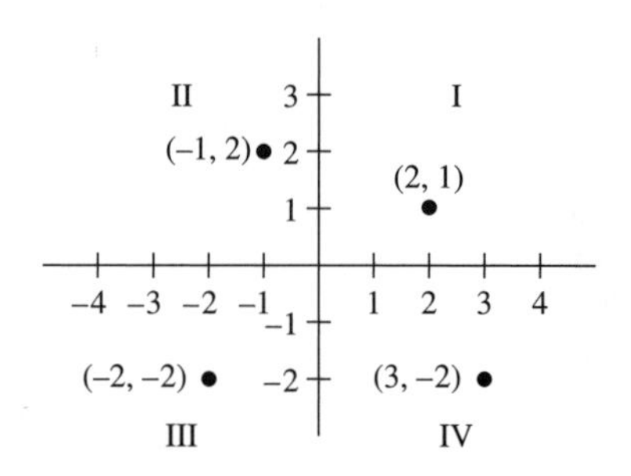

Figure 1.1

slope m goes through the y-axis at $(0, b)$, then the equation is

$$y - b = m(x - 0),$$

or

$$y = mx + b. \tag{1.2}$$

Equation (1.2) is called the **slope-intercept** form, and the number b is called the **y-intercept**.

Lines that are not parallel to the y-axis have equations of the form $y = mx + b$. Vertical lines have equations of the form $x = c$. Hence, any line has an equation of the form

$$Ax + By + C = 0, \tag{1.3}$$

where at least one of A and B is nonzero. Conversely, the graph of any equation of the form (1.3) is a line.

EXAMPLE 1.1

Write the equation of the line through $(3, -1)$ and $(1, 2)$. What is the slope, and what is the y-intercept?

Solution

We first find the slope using $m = \frac{(y_2 - y_1)}{(x_2 - x_1)}$:

$$m = \frac{2 - (-1)}{1 - 3} = -\frac{3}{2}.$$

Now use either of the given points, say $(3, -1)$, and write the equation (1.1):

$$y - (-1) = \left(-\frac{3}{2}\right)(x - 3),$$

$$y + 1 = -\frac{3}{2}x + \frac{9}{2}.$$

The slope-intercept form is

$$y = -\frac{3}{2}x + \frac{7}{2},$$

so the y-intercept is $\frac{7}{2}$, which is the value of y when $x = 0$.

EXAMPLE 1.2

Write the equation and graph the line through $(3, 1)$ with slope $\frac{1}{2}$. What are the x- and y-intercepts?

Solution

The equation of the line is

$$y - 1 = \frac{1}{2}(x - 3),$$

$$y = \frac{1}{2}x - \frac{1}{2}.$$

The y-intercept is $-\frac{1}{2}$. To find the x-intercept, where the line crosses the x-axis, set $y = 0$ and solve for x:

$$0 = \frac{1}{2}x - \frac{1}{2}; \quad x = 1.$$

The x-intercept is 1. The graph is shown in Figure 1.2.

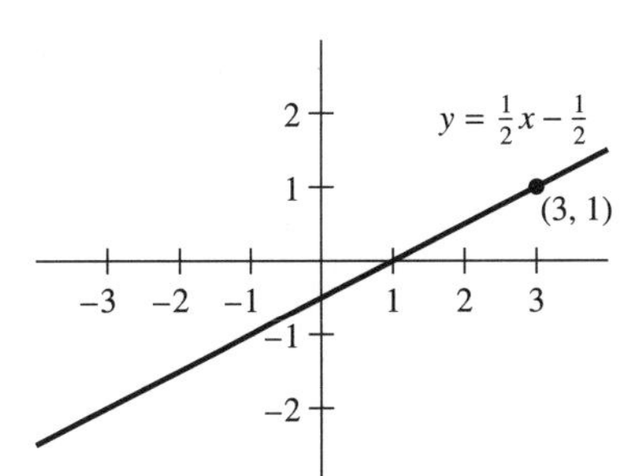

Figure 1.2

Parallel lines have the same slope, but what about perpendicular lines? Let m_1 and m_2 be the slopes of two perpendicular lines. In Figure 1.3 we see that angles BAD and BDC are equal, since their respective sides are perpendicular. Hence, $b/a = c/b$, and $b^2 = ac$. The slopes m_1 and m_2 are given by

$$m_1 = \frac{b}{a} \quad \text{and} \quad m_2 = -\frac{b}{c},$$

so

$$m_1 m_2 = \left(\frac{b}{a}\right)\left(-\frac{b}{c}\right) = -\frac{b^2}{ac} = -1.$$

Lines are perpendicular if and only if their slopes are negative reciprocals of each other.

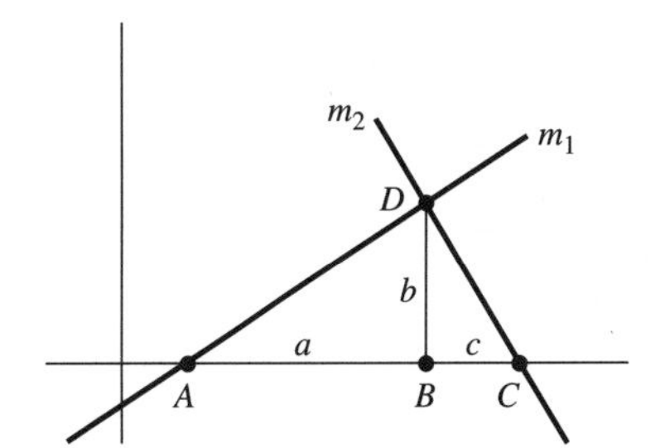

Figure 1.3

EXAMPLE 1.3

Find the line through (3, 2) which is perpendicular to the line $5x + 2y - 7 = 0$, and the line through (3, 2) which is parallel to $5x + 2y - 7 = 0$.

Solution

We write the equation of the given line in slope-intercept form to determine its slope:

$$y = -\frac{5}{2}x + \frac{7}{2}.$$

The given line has slope $-\frac{5}{2}$, so a parallel line has slope $-\frac{5}{2}$ and a perpendicular line has slope $\frac{2}{5}$. Hence, the perpendicular line through (3, 2) is

$$y - 2 = \frac{2}{5}(x - 3),$$

and the parallel line through (3, 2) is

$$(y - 2) = -\frac{5}{2}(x - 3).$$

The angle α a line makes with the x-axis, or any horizontal line, is called the **inclination** of the line. If a line has inclination α, then its slope is $m = \tan\alpha$ (Figure 1.4).

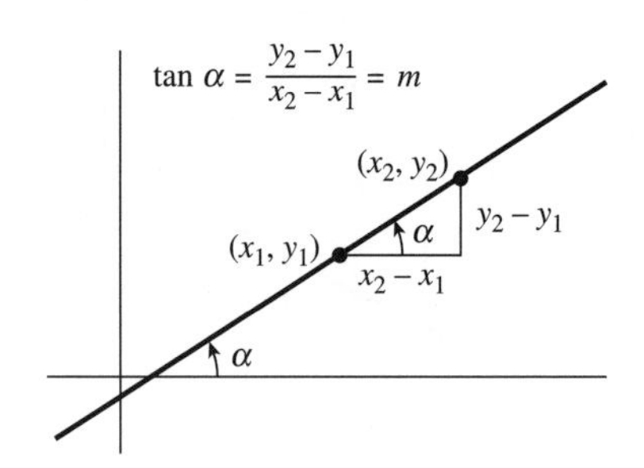

Figure 1.4

EXAMPLE 1.4

(a) If a line has inclination 20°, what is its slope?

(b) If a line has slope 3, what is its inclination?

Solution

Make sure your calculator is set for degrees and read that the slope is $\tan 20° = .36$. To find the inclination of a line with slope 3, read $\tan^{-1} 3 = 71.6°$.

EXAMPLE 1.5

Find the angle between the lines $y = x - 4$ and $y = \frac{1}{2}x - 1$ (Figure 1.5).

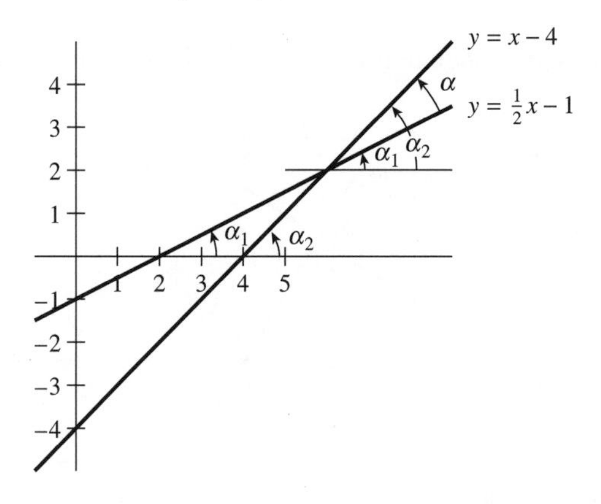

Figure 1.5

Solution

From Figure 1.5 we see that if α_2 is the inclination of $y = x - 4$, and α_1 is the inclination of $y = \frac{1}{2}x - 1$, then the angle α between the lines is $\alpha = \alpha_2 - \alpha_1$. Here

$$\alpha_2 = \tan^{-1} 1 = 45°,$$

$$\alpha_1 = \tan^{-1} \frac{1}{2} = 26.6°,$$

so $\alpha = 18.4°$.

PROBLEMS

1.1 Graph the pairs of points, and then describe the geometric relationship between any two points (a, b) and (b, a); $(1, 0)$ and $(0, 1)$; $(3, 2)$ and $(2, 3)$; $(-1, 2)$ and $(2, -1)$; $(4, -1)$ and $(-1, 4)$.

Graph the lines and determine their slopes.

1.2 $x + 2y = 4$

1.3 $x - y = 2$

1.4 $y = 2x - 3$

1.5 $y = -x - 1$

1.6 $x + 3y = 0$

1.7 $2x + 3y + 6 = 0$

Write the equations of the following lines, and put the equation in the slope-intercept form $y = mx + b$.

1.8 through (1, 1) and (3, 0)

1.9 through (−2, 3) and (4, 1)

1.10 with y-intercept 7 and slope 2

1.11 with y-intercept −1 and slope −4

1.12 through (1, 4) with slope −3

1.13 through (2, 3) and perpendicular to $x - 2y + 1 = 0$

1.14 through (1, 2) and parallel to $7x - y = 4$

1.15 (i) Show that the lines $x + y - 3 = 0$ and $x - y + 1 = 0$ intersect at (1, 2).

(ii) Show that for any constant k

$$(x - y + 1) + k(x + y - 3) = 0 \qquad (*)$$

is the equation of a line through (1, 2).

(iii) Find k so that the line (*) goes through (1, 2) and (2, 4), and write the equation of the line.

1.16 Find the equation of the line through (1, 1) and the intersection of the lines $x - 2y + 2 = 0$ and $x + y - 4 = 0$. *Hint*: See Problem 1.15. Notice that you don't have to find the intersection of the given lines.

1.17 Find the equation of the line with slope 2 which passes through the intersection of the lines $x + 2y - 5 = 0$ and $3x + y - 7 = 0$.

1.18 Find the equation of the line through the intersection of $x + 3y - 2 = 0$ and $2x + y - 5 = 0$ which is perpendicular to $x + 3y - 2 = 0$.

1.19 Find the angle (i) between the lines $y = 2x$ and $y = 3x$; (ii) between the lines $y = 2x - 5$ and $y = 3x + 7$.

1.20 Find the line through (3, 4) which makes an angle of 25° with the x-axis.

2

Parabolas, Ellipses, Hyperbolas

Mathematicians in ancient Greece discovered that if you cut a cone with a plane, you get curves with very interesting properties. For example, if the plane is perpendicular to the axis of the cone, you get a circle. Tilt the plane a little and the circle becomes an ellipse. Keep tilting and at one critical angle the curve is a parabola. Any further tilts yield a hyperbola. These curves are still of basic interest, and we will encounter them frequently. In the coordinate plane these curves, called conic sections for the obvious reason, all have equations of the second degree. We will consider the following simple cases:

$$y = Ax^2 + Bx + C, \quad \text{parabolas;}$$

$$\frac{x^2}{a^2} + \frac{y^2}{b^2} = 1, \quad \text{ellipses;}$$

$$\frac{x^2}{a^2} - \frac{y^2}{b^2} = \pm 1, \quad \text{hyperbolas.}$$

Consider first the parabola $y = x^2$. The graph is symmetric about the y-axis since $(-x, y)$ lies on the curve whenever (x, y) does. The graph is shown in Figure 2.1.

If the coefficient of x^2 is larger (e.g., $y = 2x^2$), then the curve heads up more sharply, and if the coefficient is negative (e.g., $y = -\frac{1}{2}x^2$), the curve heads downward as $|x|$ increases.

All the curves $y = Ax^2 + Bx + C$ with $A \neq 0$ are parabolas, and they all have exactly the same shape as $y = Ax^2$. The axis of the parabola will move right or left depending on B, and the curve will move up or down depending on B and C, but the shape remains the same as $y = Ax^2$. Consider, for example, $y = x^2 - 4x + 3$. We show that this curve has the same shape as $y = x^2$ by completing the square:

$$y = x^2 - 4x + 3,$$

$$y = x^2 - 4x + 4 - 1,$$

$$y = (x - 2)^2 - 1.$$

The graph of $y = (x - 2)^2$ is just the graph of $y = x^2$ moved over so that its axis is the line $x = 2$. The constant -1 drops the whole curve down one unit. The three curves $y = x^2$, $y = (x - 2)^2$, and $y = (x - 2)^2 - 1$ are shown in Figure 2.2.

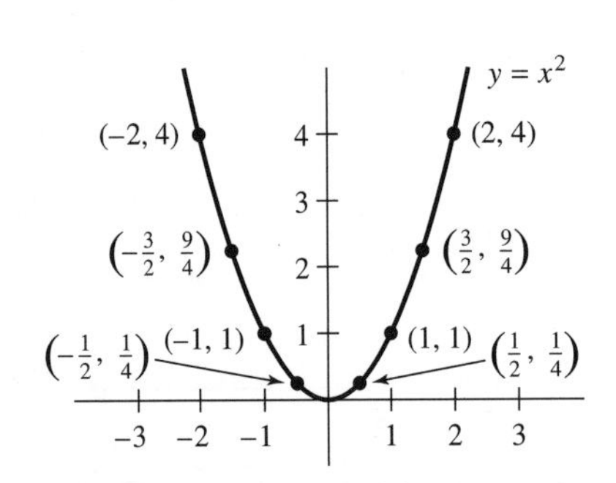

Figure 2.1

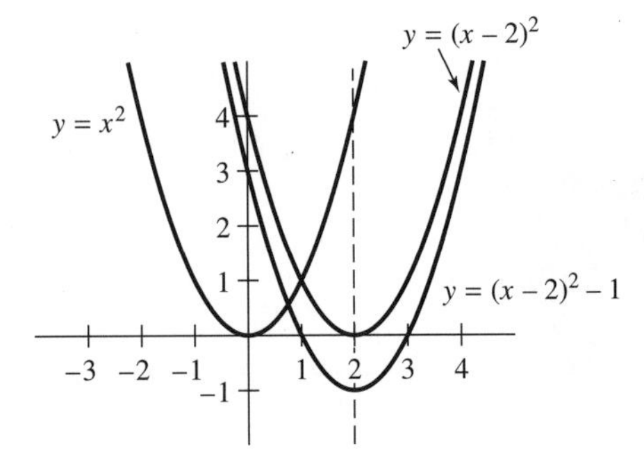

Figure 2.2

The distance r between points (x_1, y_1) and (x_2, y_2) is given by

$$r = \sqrt{(x_2 - x_1)^2 + (y_2 - y_1)^2}.$$

It follows that the circle with center (x_0, y_0) and radius r is characterized by the equation

$$\sqrt{(x - x_0) + (y - y_0)^2} = r,$$

or

$$(x - x_0)^2 + (y - y_0)^2 = r^2. \tag{2.1}$$

From (2.1) we see that every circle has an equation of the form

$$x^2 + y^2 + Ax + By + C = 0. \tag{2.2}$$

Conversely, every equation of form (2.2) which has a graph represents a circle or just a single point. For example, $x^2 + y^2 + 1 = 0$ has no graph, and the graph of $x^2 + y^2 = 0$ is just the single point $(0, 0)$.

EXAMPLE 2.1

Find the center and radius of the circle $x^2 + y^2 - 2x + 4y - 4 = 0$.

Solution

We complete the squares to put the equation in the form (2.1):

$$x^2 - 2x + y^2 + 4y = 4,$$
$$x^2 - 2x + 1 + y^2 + 4y + 4 = 4 + 1 + 4,$$
$$(x - 1)^2 + (y + 2)^2 = 9.$$

The circle has center $(1, -2)$ and radius 3.

The graph of

$$\frac{x^2}{a^2} + \frac{y^2}{b^2} = 1$$

is an ellipse through the points $(\pm a, 0)$ and $(0, \pm b)$ (Figure 2.3). If $a = b$, then the ellipse is a circle of radius a. If $b > a$, then the ellipse is longer in the y-direction.

The curves

$$\frac{x^2}{a^2} - \frac{y^2}{b^2} = 1 \quad \text{and} \quad -\frac{x^2}{a^2} + \frac{y^2}{b^2} = 1$$

are hyperbolas. The first hyperbola above intersects the x-axis, and the second intersects the y-axis (Figure 2.4). Both hyperbolas approach the **asymptotes** $y = \pm\frac{b}{a}x$ as x increases; that is, the vertical distance between the curve and the asymptote tends to zero as x increases.

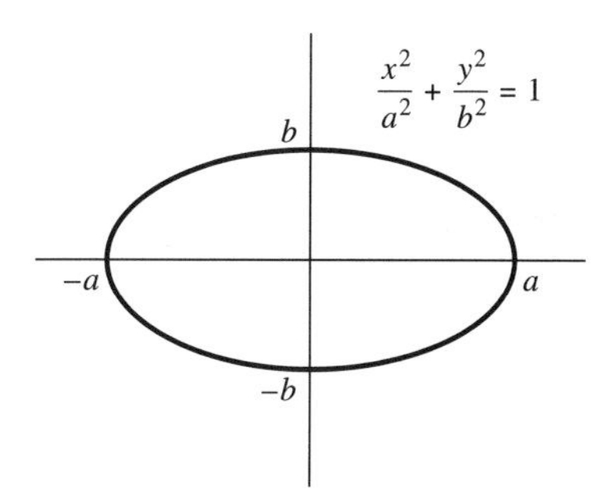

Figure 2.3

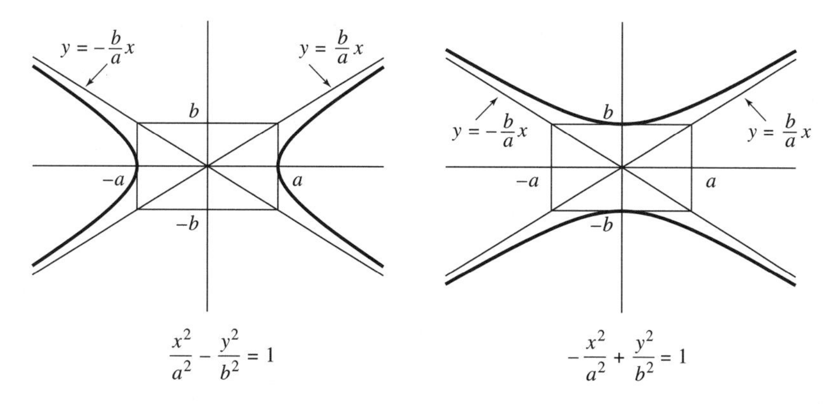

Figure 2.4

EXAMPLE 2.2

Graph the curve $-x^2 + 4y^2 = 4$.

Solution

If we divide both sides by 4, we recognize the equation of a hyperbola:

$$-\frac{x^2}{4} + \frac{y^2}{1} = 1.$$

The hyperbola intersects the y-axis at $(0, \pm 1)$ and has the lines $y = \pm\frac{1}{2}x$ as its asymptotes (Figure 2.5).

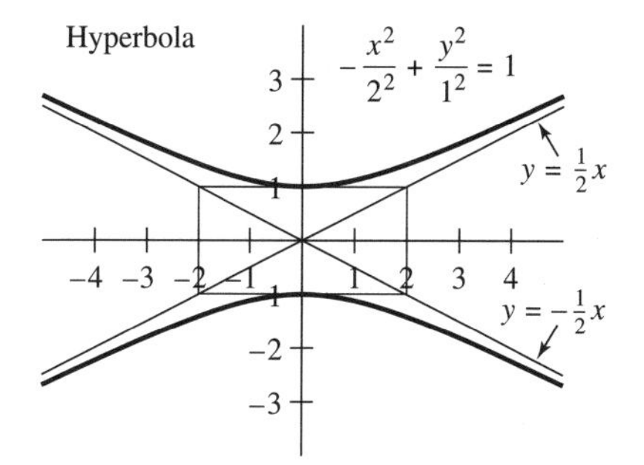

Figure 2.5

The equation $y = \frac{1}{x}$ or $xy = 1$ (Figure 2.6) also turns out to be a hyperbola. Specifically, this is the hyperbola $-\frac{x^2}{2} + \frac{y^2}{2} = 1$ rotated through 45° in the clockwise direction. The asymptotes are the coordinate axes. The graph of $y = \frac{1}{x-2}$ is the graph of $y = \frac{1}{x}$ moved two units to the right, and the graphs of the curves $y = \frac{A}{x-a}$ are similar, with constant A effecting a scaling in the y-direction.

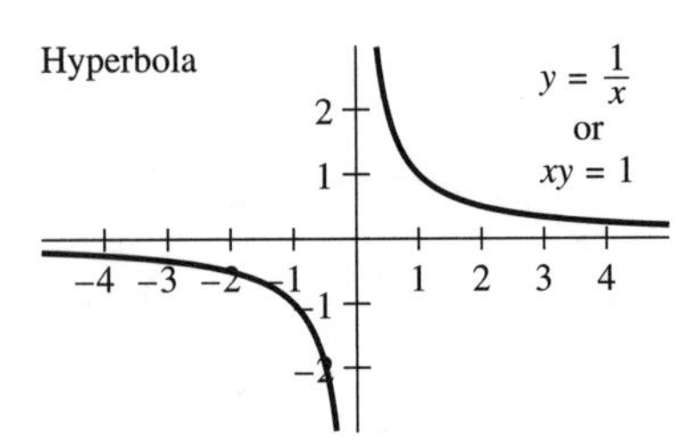

Figure 2.6

EXAMPLE 2.3

Find the equation of the circle which passes through the origin and the two points where the line $x+y=1$ intersects the circle $x^2+y^2=1$.

Solution

Any circle has the form $x^2+y^2+Ax+By+C=0$, so three given points will determine the three constants A, B, C. One solution would be to find the two points where $x+y=1$ and $x^2+y^2=1$ intersect and use these two points and (0, 0) to write three equations in A, B, and C. Here is an easier way: Any equation

$$(x^2+y^2-1)+k(x+y-1)=0 \qquad (*)$$

is a circle because it has the right form. If a point (x_0, y_0) satisfies both $x^2+y^2-1=0$ and $x+y-1=0$, then it will certainly lie on the circle (*). We find k so the circle (*) also goes through (0, 0),

$$(-0^2+0^2-1)+k(0+0-1),$$
$$k=-1.$$

Therefore, the sought-for circle is

$$x^2+y^2-1-(x+y-1)=0,$$

or

$$x^2+y^2-x-y=0.$$

PROBLEMS

Graph the following curves.

2.1 $y=-x^2$

2.2 $y=2x^2$

2.3 $y=\frac{1}{2}x^2$

2.4 $y=(x+1)^2$

2.5 $y=x^2+1$

2.6 $y=1-x^2$

2.7 $y=x^2-2x+2$

2.8 $y=-x^2+2x$

2.9 $x^2+y^2=4$

2.10 $(x-1)^2+y^2=1$

2.11 $x^2+y^2-4y-5=0$

2.12 $\frac{x^2}{9}+\frac{y^2}{4}=1$

2.13 $x^2+\frac{y^2}{4}=1$

2.14 $4x^2+9y^2=36$

2.15 $x^2 - y^2 = 1$

2.16 $y^2 - x^2 = 1$

2.17 $xy = 2$

2.18 $y = \dfrac{1}{x-1}$

2.19 Find the equation of the circle which passes through the origin and the two points where $y = \frac{1}{2}x - 1$ intersects the circle $x^2 + y^2 = 4$.

2.20 Find the parabola through the point $(-4, 3)$ and the two points where the line $y = x + 1$ intersects the parabola $y = x^2 - 1$.

2.21 Show that the set of points (x, y) such that the distance from (x, y) to $(c, 0)$ plus the distance from (x, y) to $(-c, 0)$ equals $2a$ is the ellipse $\frac{x^2}{a^2} + \frac{y^2}{a^2-c^2} = 1$. *Hint*: Put the square roots on opposite sides of the equation and square both sides. Isolate the remaining square root and square both sides again.

2.22 Show that the top part of the hyperbola $\frac{x^2}{a^2} - \frac{y^2}{b^2} = 1$ (i.e., the curve $y = b\sqrt{\frac{x^2}{a^2} - 1} = \frac{b}{a}\sqrt{x^2 - a^2}$) approaches the asymptote $y = \frac{b}{a}x$ as $x \longrightarrow \infty$. *Hint*: To show $x - \sqrt{x^2 - a^2} \longrightarrow 0$, multiply and divide by $x + \sqrt{x^2 - a^2}$ and let $x \longrightarrow \infty$.

3

Differentiation

If $s(t)$ is the position at time t of a car moving along a straight road, then the average speed over the distance from $s(t_0)$ to $s(t)$ is the distance traveled divided by the elapsed time:

$$\text{average speed} = \frac{s(t) - s(t_0)}{t - t_0}.$$

If the speed of the car is not constant, then the average speed may have little to do with the actual speed at any given time. However, if the time interval $[t_0, t]$ is very small, then the speed will not vary much over the interval, and the actual speed at any given time will be close to the average speed. We **define** the speed at a particular time t_0, the **instantaneous speed** at t_0, to be the limit of the average speeds over smaller and smaller time intervals $[t_0, t]$. This limit is denoted $s'(t_0)$, and we write

$$s'(t_0) = \lim_{t \to t_0} \frac{s(t) - s(t_0)}{t - t_0}. \tag{3.1}$$

The number $s'(t_0)$ is called the **derivative of s with respect to t at t_0**, or the **rate of change of s with respect to t at t_0**.

EXAMPLE 3.1

The distance s in feet which a falling object travels in t seconds is approximately $s = 16t^2$. (This formula holds for the first few seconds, and then air resistance takes over. Falling bodies do not make sonic booms.) The distance traveled between $t_0 = 2$ seconds and a subsequent time t is

$$\begin{aligned} s(t) - s(2) &= 16t^2 - 16 \cdot 2^2 \\ &= 16(t^2 - 4) \\ &= 16(t + 2)(t - 2). \end{aligned}$$

Hence, the average speed over the time interval $[2, t]$ is

$$\frac{s(t) - s(2)}{t - 2} = \frac{16(t + 2)(t - 2)}{t - 2} = 16(t + 2),$$

and the instantaneous speed at $t = 2$ is

$$s'(2) = \lim_{t \to 2} \frac{s(t) - s(2)}{t - 2}$$
$$= \lim_{t \to 2} 16(t + 2) = 64.$$

Since s is measured in feet and t in seconds, the speed at $t = 2$ is 64 ft/sec.

Now consider what the derivative of a function means in a purely geometric setting. We look at the function $y = f(x) = \frac{1}{2}x^2$, whose graph is the parabola shown in Figure 3.1. The change in y over the interval from 1 to a nearby point x is

$$f(x) - f(1) = \frac{1}{2}x^2 - \frac{1}{2} \cdot 1 = \frac{1}{2}(x + 1)(x - 1).$$

Hence, the average rate of change of y over this interval is

$$\frac{f(x) - f(1)}{x - 1} = \frac{\frac{1}{2}(x + 1)(x - 1)}{x - 1} = \frac{1}{2}(x + 1).$$

The above quotient $(f(x) - f(1))/(x - 1)$ is the slope $m(x)$ of the secant line joining the two points $(1, \frac{1}{2})$ and $(x, \frac{1}{2}x^2)$. The derivative, $f'(1)$, is therefore the limiting slope of these secant lines:

$$f'(1) = \lim_{x \to 1} m(x)$$
$$= \lim_{x \to 1} \frac{1}{2}(x + 1) = 1.$$

We *define* the tangent line to the curve at $(1, \frac{1}{2})$ to be the line that has this limiting slope 1. Hence, the tangent line to $y = \frac{1}{2}x^2$ at $(1, \frac{1}{2})$ is

$$y - \frac{1}{2} = 1 \cdot (x - 1).$$

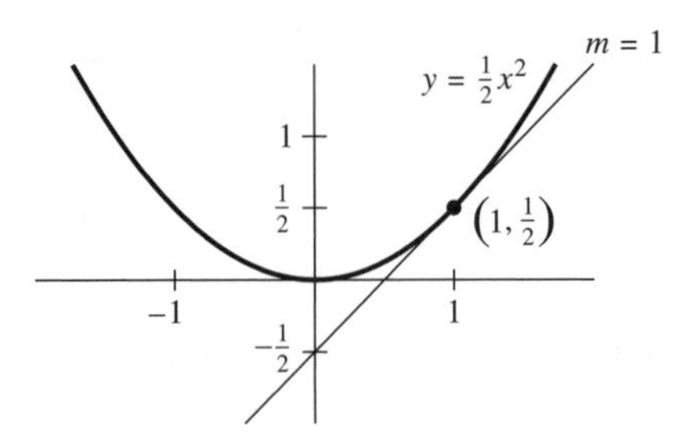

Figure 3.1

Now we have these two ways to interpret the derivative

$$f'(x_0) = \lim_{x \to x_0} \frac{f(x) - f(x_0)}{x - x_0}: \tag{3.2}$$

$f'(x_0)$ is the rate of change of f with respect to x (or of y with respect to x if $y = f(x)$), and is also the slope of the tangent line to $y = f(x)$ at x_0. The expression $(f(x) - f(x_0))/(x - x_0)$ is the **difference quotient for** f **at** x_0.

In the preceding examples, we used the following two obvious limits:

$$\lim_{t \to 2} 16(t + 2) = 64; \ \lim_{x \to 1} \frac{1}{2}(x + 1) = 1.$$

The meaning of the limit concept is as follows: a function $q(x)$ (such as the difference quotient in (3.2)) approaches a limit L as x approaches x_0, written $\lim_{x \to x_0} q(x) = L$, provided that $q(x)$ is arbitrarily close to L for all x sufficiently close to x_0. The phrases "arbitrarily close" and "sufficiently close" can be made arithmetically precise as follows: $\lim_{x \to x_0} q(x) = L$ provided that for any given positive number ε (an arbitrary choice of closeness) there corresponds a positive number δ (this is how close "sufficiently close" is) such that $q(x)$ is within ε of L if x is within δ of x_0. The intuitive idea of $q(x)$ approaching L as a limiting value as x approaches x_0 will suffice for the limits we deal with in this course.

The following examples show some more limits in action, and the action is again the calculation of derivatives.

EXAMPLE 3.2

Find $f'(-1)$ if $f(x) = 3x^2$.

Solution

$$\begin{aligned} f'(-1) &= \lim_{x \to -1} \frac{f(x) - f(-1)}{x - (-1)} \\ &= \lim_{x \to -1} \frac{3x^2 - 3 \cdot 1}{x + 1} \\ &= \lim_{x \to -1} \frac{3(x+1)(x-1)}{x+1} \\ &= \lim_{x \to -1} 3(x - 1) = -6. \end{aligned}$$

EXAMPLE 3.3

Find $f'(4)$ if $f(x) = \sqrt{x}$.

Solution

$$\begin{aligned} f'(4) &= \lim_{x \to 4} \frac{f(x) - f(4)}{x - 4} \\ &= \lim_{x \to 4} \frac{\sqrt{x} - \sqrt{4}}{x - 4} \\ &= \lim_{x \to 4} \frac{(\sqrt{x} - 2)(\sqrt{x} + 2)}{(x - 4)(\sqrt{x} + 2)} \\ &= \lim_{x \to 4} \frac{x - 4}{(x - 4)(\sqrt{x} + 2)} \\ &= \lim_{x \to 4} \frac{1}{\sqrt{x} + 2} = \frac{1}{4}. \end{aligned}$$

EXAMPLE 3.4

Find the equation of the line tangent to $y = \frac{3}{x}$ at $x = -2$.

Solution

The slope we want is $f'(-2)$ where $f(x) = \frac{3}{x}$.

$$\begin{aligned} f'(-2) &= \lim_{x \to -2} \frac{\frac{3}{x} - \frac{3}{-2}}{x - (-2)} \\ &= \lim_{x \to -2} \frac{3\left(\frac{1}{x} + \frac{1}{2}\right)}{x + 2} \\ &= \lim_{x \to -2} \frac{3(2 + x)}{2x(x + 2)} \\ &= \lim_{x \to -2} \frac{3}{2x} = -\frac{3}{4}. \end{aligned}$$

The equation of the tangent line at $(-2, -\frac{3}{2})$ is

$$y + \frac{3}{2} = \left(-\frac{3}{4}\right)(x + 2).$$

EXAMPLE 3.5

Find $f'(4)$ if $f(x) = \sqrt{x} + \frac{1}{x}$.

Solution

$$\begin{aligned}
f'(4) &= \lim_{x \to 4} \frac{\sqrt{x} + \frac{1}{x} - \left(\sqrt{4} + \frac{1}{4}\right)}{x - 4} \\
&= \lim_{x \to 4} \frac{(\sqrt{x} - \sqrt{4}) + (\frac{1}{x} - \frac{1}{4})}{x - 4} \\
&= \lim_{x \to 4} \left(\frac{(\sqrt{x} - \sqrt{4})(\sqrt{x} + \sqrt{4})}{(x - 4)(\sqrt{x} + \sqrt{4})} + \frac{4 - x}{4x(x - 4)} \right) \\
&= \lim_{x \to 4} \left(\frac{x - 4}{(x - 4)(\sqrt{x} + \sqrt{4})} - \frac{1}{4x} \right) \\
&= \frac{1}{4} - \frac{1}{16} = \frac{3}{16}.
\end{aligned}$$

Notice that in Example 3.5 we effectively found the derivatives of $\sqrt{x}$ and $\frac{1}{x}$ separately and then added the results. That is because the difference quotient for $\sqrt{x} + \frac{1}{x}$ can be separated into two groups, one group being the difference quotient for $\sqrt{x}$ and the other being the difference quotient for $\frac{1}{x}$.

EXAMPLE 3.6

A ball thrown downward with an initial speed of 32 ft/sec from the top of a tall building travels a vertical distance of s feet in t seconds, where $s = 16t^2 + 32t$. What is its speed at $t = 1$?

Solution

We are asked to find $s'(1)$, and again we separate the difference quotient into the difference quotient for $16t^2$ and that for $32t$.

$$\begin{aligned}
s'(1) &= \lim_{t \to 1} \frac{s(t) - s(1)}{t - 1} \\
&= \lim_{t \to 1} \frac{16t^2 + 32t - (16 + 32)}{t - 1} \\
&= \lim_{t \to 1} \left[\frac{16t^2 - 16}{t - 1} + \frac{32t - 32}{t - 1} \right] \\
&= \lim_{t \to 1} \left[\frac{16(t + 1)(t - 1)}{t - 1} + \frac{32(t - 1)}{t - 1} \right] \\
&= \lim_{t \to 1} [16(t + 1) + 32] = 64.
\end{aligned}$$

PROBLEMS

Find $f'(a)$ for the specified a.

3.1 $f(x) = 2x^2;\ a = 1$

3.2 $f(x) = -5x^2;\ a = -2$

3.3 $f(x) = x^3$; $a = 3$
Hint: $x^3 - a^3 = (x - a)(x^2 + ax + a^2)$.

3.4 $f(x) = 2x^3$; $a = 1$

3.5 $f(x) = 3\sqrt{x}$; $a = 9$

3.6 $f(x) = -\sqrt{x} + x$; $a = 1$

3.7 $f(x) = \dfrac{1}{x}$; $a = 3$

3.8 $f(x) = \dfrac{1}{x^2}$; $a = 1$

3.9 $f(x) = x^{\frac{1}{3}}$; $a = 8$
Hint: $(x^{\frac{1}{3}} - a^{\frac{1}{3}})(x^{\frac{2}{3}} + a^{\frac{1}{3}}x^{\frac{1}{3}} + a^{\frac{2}{3}}) = x - a$.

Find the equation of the line tangent to the curve at the given point.

3.10 $y = 2x^2$; $(1, 2)$

3.11 $y = 5 - x^2$; $(2, 1)$

3.12 $y = 2\sqrt{x}$; $(4, 4)$

3.13 $y = \dfrac{1}{\sqrt{x}}$; $(1, 1)$

3.14 A ball is thrown straight up from the ground, and its height s in feet after t seconds is $s = 100t - 16t^2$. What is the speed of the ball at $t = 2$ seconds? What is the initial speed $s'(0)$? How high does the ball go? (*Hint*: The ball reaches maximum height when its speed is zero.)

3.15 A ball thrown down from the top of a 104-ft building, with an initial speed of 20 ft/sec, travels a distance $s = 20t + 16t^2$ in t seconds. When does it hit the ground, and how fast is it going then?

3.16 A spherical balloon is blown up in such a way that its radius at t seconds is t cm; that is, $r = t$. What is the rate of change of the volume at any time t? The volume is $V = \frac{4}{3}\pi r^3$, and the surface area is $S = 4\pi r^2$. How do you interpret the fact that $\frac{dV}{dt} = S$?

3.17 A block of ice initially weighs 100 lbs and melts so that its weight W after t minutes is $100 - 2\sqrt{t}$. At what rate (lbs/min) is the ice melting at $t = 100$ min?

4

Differentiation Formulas

In this section we develop some general rules for differentiating functions, so we don't have to go through the limit argument each time. What we have so far are techniques that will give us the derivative of a specific function at a specific point—for example, the derivative of x^2 at $x_0 = 1$. Now we search for a formula for the derivative of x^2 at *any* x, and of course we want such formulas for all the common functions.

Our definition for $f'(x_0)$ is the following:

$$f'(x_0) = \lim_{x \to x_0} \frac{f(x) - f(x_0)}{x - x_0}. \tag{4.1}$$

To get a formula for $f'(x)$, for any x, we replace x_0 in (4.1) by x, and write the nearby point as $x + \Delta x$. Here Δx just stands for a small change in x. With this notation, $f(x) - f(x_0)$ becomes $f(x + \Delta x) - f(x)$, and $x - x_0$ becomes Δx. This gives us the alternative form

$$f'(x) = \lim_{\Delta x \to 0} \frac{f(x + \Delta x) - f(x)}{\Delta x}. \tag{4.2}$$

If $y = f(x)$, we let $\Delta y = f(x + \Delta x) - f(x)$, so Δy is the change in y that corresponds to the change Δx in x. The difference quotient at any point x is

$$\frac{\Delta y}{\Delta x} = \frac{f(x + \Delta x) - f(x)}{\Delta x}.$$

We write $\frac{dy}{dx}$ for the limit of $\frac{\Delta y}{\Delta x}$ as $\Delta x \longrightarrow 0$, so we have the following alternative notations:

$$\frac{dy}{dx} = \lim_{\Delta x \to 0} \frac{\Delta y}{\Delta x} = \lim_{\Delta x \to 0} \frac{f(x + \Delta x) - f(x)}{\Delta x} = f'(x).$$

We also use $\frac{d}{dx}$ as a differentiation operator so that $\frac{d}{dx} f(x) = f'(x)$.

In the examples of the last section, we saw that to differentiate the sum of two terms you just differentiate the terms separately and add the results. Thus, if $y = z + w$, where z and w are functions of x, then

$$\Delta y = \Delta z + \Delta w,$$

and consequently

$$\begin{aligned}\frac{\Delta y}{\Delta x} &= \frac{\Delta z}{\Delta x} + \frac{\Delta w}{\Delta x},\\ \frac{dy}{dx} &= \lim_{\Delta x \to 0}\left(\frac{\Delta z}{\Delta x} + \frac{\Delta w}{\Delta x}\right)\\ &= \lim_{\Delta x \to 0}\frac{\Delta z}{\Delta x} + \lim_{\Delta x \to 0}\frac{\Delta w}{\Delta x}\\ &= \frac{dz}{dx} + \frac{dw}{dx}.\end{aligned}$$

The derivative of a sum (or difference) is the sum (or difference) of the derivatives. In the second equation above we used the fact that the limit of a sum is the sum of the limits. This is an obvious property of addition: If a is close to A and b is close to B, then $a + b$ is close to $A + B$. Similar statements hold for $a - b$, ab, and a/b, so, for example, the limit of a product is the product of the limits, and so on.

Now consider the derivative of a product of two functions. Let $y = z \cdot w$, and again let Δz, Δw be the changes in z and w that result when x is changed by an amount Δx. The new value of y is $(z + \Delta z)(w + \Delta w)$, and therefore

$$\begin{aligned}\Delta y &= (z + \Delta z)(w + \Delta w) - zw\\ &= z \cdot \Delta w + w \cdot \Delta z + \Delta z \cdot \Delta w.\end{aligned}$$

It follows that

$$\begin{aligned}\frac{dy}{dx} &= \lim_{\Delta x \to 0}\left(z\frac{\Delta w}{\Delta x} + w\frac{\Delta z}{\Delta x} + \Delta z \cdot \frac{\Delta w}{\Delta x}\right)\\ &= z\frac{dw}{dx} + w\frac{dz}{dx} + 0 \cdot \frac{dw}{dx} \qquad (4.3)\\ &= z\frac{dw}{dx} + w\frac{dz}{dx}.\end{aligned}$$

Of course $\lim_{\Delta x \to 0} \Delta z = 0$, for otherwise $\Delta z/\Delta x$ would not approach a limit as $\Delta x \longrightarrow 0$. A function z with the property that $\Delta z \longrightarrow 0$ as $\Delta x \longrightarrow 0$ is **continuous** at the point in question. Differentiable functions are continuous, and calculus deals with differentiable functions.

In functional notation formula (4.3) can be written as follows: if $f(x) = g(x) \cdot h(x)$, then

$$f'(x) = g(x)h'(x) + h(x)g'(x),$$

or

$$\frac{d}{dx}f(x) = g(x)\frac{d}{dx}h(x) + h(x)\frac{d}{dx}g(x).$$

The derivative (rate of change) of a constant function is obviously zero: $\frac{dc}{dx} = 0$. It is also clear from the definition that

$$\frac{d}{dx}(cy) = c\frac{dy}{dx}.$$

The derivative of the function x is clearly one ($\frac{dx}{dx} = 1$), so $\frac{d}{dx}(cx) = c$.

Now we use the product rule to find a formula for $\frac{d}{dx}x^n$.

$$\frac{d}{dx}x^2 = \frac{d}{dx}(x \cdot x) = x\frac{dx}{dx} + x\frac{dx}{dx} = 2x,$$

$$\frac{d}{dx}x^3 = \frac{d}{dx}(x \cdot x^2) = x\frac{d}{dx}x^2 + x^2\frac{dx}{dx} = x \cdot 2x + x^2 = 3x^2,$$

$$\frac{d}{dx}x^4 = \frac{d}{dx}(x \cdot x^3) = x\frac{d}{dx}x^3 + x^3\frac{dx}{dx} = x \cdot 3x^2 + x^3 = 4x^3.$$

The pattern is clear:

$$\frac{d}{dx}x^n = nx^{n-1}, \quad n = 0, 1, 2, 3, \ldots. \tag{4.4}$$

It follows that

$$\frac{d}{dx}(cx^n) = cnx^{n-1}.$$

EXAMPLE 4.1

Find $\frac{dy}{dx}$ if $y = 5x^{10} - 3x^5 + 2x^2 - 5$.

Solution

Differentiating the terms separately and adding, we get

$$\frac{dy}{dx} = 50x^9 - 15x^4 + 4x.$$

To find a formula for $\frac{d}{dx}x^{-n}$, we start with x^{-1} and find $\frac{dx^{-1}}{dx}$ from the definition:

$$\begin{aligned}\frac{d}{dx}\frac{1}{x} &= \lim_{\Delta x \to 0} \frac{\frac{1}{x+\Delta x} - \frac{1}{x}}{\Delta x} \\ &= \lim_{\Delta x \to 0} \frac{x - (x + \Delta x)}{\Delta x \cdot x(x + \Delta x)} \\ &= \lim_{\Delta x \to 0} \frac{-1}{x(x + \Delta x)} \\ &= \frac{-1}{x^2}.\end{aligned}$$

That is,

$$\frac{d}{dx}x^{-1} = -x^{-2}. \tag{4.5}$$

Now proceed as before, using the product rule and (4.5):

$$\begin{aligned}\frac{d}{dx}x^{-2} &= \frac{d}{dx}(x^{-1} \cdot x^{-1}) = x^{-1}\frac{dx^{-1}}{dx} + x^{-1}\frac{dx^{-1}}{dx} \\ &= x^{-1}(-x^{-2}) + x^{-1}(-x^{-2}) = -2x^{-3},\end{aligned}$$

$$\begin{aligned}\frac{d}{dx}x^{-3} &= \frac{d}{dx}(x^{-1} \cdot x^{-2}) = x^{-1}\frac{dx^{-2}}{dx} + x^{-2}\frac{dx^{-1}}{dx} \\ &= x^{-1}(-2x^{-3}) + x^{-2}(-x^{-2}) = -3x^{-4}.\end{aligned}$$

Similarly, we show that

$$\frac{d}{dx}x^{-4} = -4x^{-5}, \quad \frac{d}{dx}x^{-5} = -5x^{-6},$$

and so on, and the general formula is

$$\frac{d}{dx}x^{-n} = -nx^{-n-1}, \quad n = 1, 2, 3, \ldots. \tag{4.6}$$

Notice that the rule in (4.6) is the same as that for positive exponents: to differentiate x^n, for n *positive or negative*, multiply by the exponent, and subtract one from the exponent to get the new exponent.

EXAMPLE 4.2

Find $\frac{dy}{dx}$ if $y = \frac{5}{x^3} - \frac{7}{x^2} + 2x^4$.

Solution

Write the expression with negative exponents where appropriate, and use (4.4) and (4.6).

$$y = 5x^{-3} - 7x^{-2} + 2x^4,$$

$$\frac{dy}{dx} = -15x^{-4} + 14x^{-3} + 8x^3.$$

To find a formula for the derivative of the quotient of two functions, $\frac{d}{dx}(\frac{z}{w})$, we first find $\frac{d}{dx}(\frac{1}{w})$, and then we can use the product rule to find $\frac{d}{dx}(\frac{z}{w}) = \frac{d}{dx}(z \cdot \frac{1}{w})$. If $y = \frac{1}{w}$, then

$$\begin{aligned}
\frac{\Delta y}{\Delta x} &= \frac{1}{\Delta x}\left[\frac{1}{w + \Delta w} - \frac{1}{w}\right] \\
&= \frac{1}{\Delta x}\left[\frac{w - (w + \Delta w)}{w(w + \Delta x)}\right] \\
&= \frac{-1}{w(w + \Delta x)} \cdot \frac{\Delta w}{\Delta x}.
\end{aligned}$$

Hence, taking the limit as $\Delta x \longrightarrow 0$,

$$\frac{dy}{dx} = \frac{d}{dx}\left(\frac{1}{w}\right) = -\frac{1}{w^2}\frac{dw}{dx}. \tag{4.7}$$

Now we use the product rule to find $\frac{d}{dx}(\frac{z}{w})$:

$$\begin{aligned}
\frac{d}{dx}\left(\frac{z}{w}\right) &= \frac{d}{dx}\left(z \cdot \frac{1}{w}\right) \\
&= z\frac{d}{dx}\left(\frac{1}{w}\right) + \frac{1}{w}\frac{dz}{dx} \\
&= z\left(\frac{-1}{w^2}\right)\frac{dw}{dx} + \frac{1}{w}\frac{dz}{dx} \\
&= \frac{-z\frac{dw}{dx} + w\frac{dz}{dx}}{w^2} \\
&= \frac{w\frac{dz}{dx} - z\frac{dw}{dx}}{w^2}.
\end{aligned} \tag{4.8}$$

This formula is best remembered as "the bottom times the derivative of the top, minus the top times the derivative of the bottom, over the bottom squared," for this chant doesn't depend on which variables are used for the numerator and denominator.

EXAMPLE 4.3

Find $\frac{dy}{dx}$ if $y = \frac{x^3+3x}{x^2+2}$.

Solution

$$\begin{aligned}\frac{dy}{dx} &= \frac{(x^2+2)\frac{d}{dx}(x^3+3x)-(x^3+3x)\frac{d}{dx}(x^2+2)}{(x^2+2)^2}\\ &= \frac{(x^2+2)(3x^2+3)-(x^3+3x)2x}{(x+2)^2}\\ &= \frac{3x^4+9x^2+6-2x^4-6x^2}{(x+2)^2}\\ &= (x^4+3x^2+6)/(x+2)^2.\end{aligned}$$

EXAMPLE 4.4

Find $\frac{dy}{dx}$ if $y=(x^2+1)(4x-1)$.

Solution

We can either use the product rule, or first multiply the two terms and differentiate the resulting sum. With the product rule we get

$$\begin{aligned}\frac{dy}{dx} &= (x^2+1)\cdot 4+(4x-1)2x\\ &= 4x^2+4+8x^2-2x\\ &= 12x^2-2x+4.\end{aligned}$$

If we multiply out first, we have

$$\begin{aligned}y &= (x^2+1)(4x-1)=4x^3-x^2+4x-1,\\ \frac{dy}{dx} &= 12x^2-2x+4.\end{aligned}$$

To add some variety to our repertoire, we now compute $\frac{d}{dx}\sqrt{x}$:

$$\begin{aligned}\frac{d}{dx}\sqrt{x} &= \lim_{\Delta x\to 0}\frac{\sqrt{x+\Delta x}-\sqrt{x}}{\Delta x}\\ &= \lim_{\Delta x\to 0}\frac{(\sqrt{x+\Delta x}-\sqrt{x})(\sqrt{x+\Delta x}+\sqrt{x})}{\Delta x(\sqrt{x+\Delta x}+\sqrt{x})}\\ &= \lim_{\Delta x\to 0}\frac{x+\Delta x-x}{\Delta x\left(\sqrt{x+\Delta x}+\sqrt{x}\right)}\\ &= \frac{1}{2\sqrt{x}}=\frac{1}{2}x^{-\frac{1}{2}}.\end{aligned} \tag{4.9}$$

Notice that the formula above, $\frac{d}{dx}x^{\frac{1}{2}}=\frac{1}{2}x^{-\frac{1}{2}}$, follows the same rule as for integer exponents: $\frac{d}{dx}x^n=nx^{n-1}$.

EXAMPLE 4.5

Find $\frac{d}{dx}\frac{\sqrt{x}}{3x-2}$.

Solution

$$\begin{aligned}\frac{d}{dx}\frac{\sqrt{x}}{3x-2} &= \frac{(3x-2)\frac{1}{2\sqrt{x}}-\sqrt{x}\cdot 3}{(3x-2)^2}\\ &= \frac{3x-2-6x}{2\sqrt{x}(3x-2)^2}\\ &= -\frac{3x+2}{2\sqrt{x}(3x-2)^2}.\end{aligned}$$

PROBLEMS

Find $\frac{dy}{dx}$.

4.1 $y = 5x^4 + x^2 + x$

4.2 $y = -3x^2 + \frac{1}{2}x - 2$

4.3 $y = 3x^{-4} - x^{-1} + 2x^3$

4.4 $y = \frac{2}{x^4} - \frac{3}{x^3} + 5x + \frac{1}{2}x^6$

4.5 $y = x^3 + \frac{1}{x^3}$

4.6 $y = x^{10} - \frac{1}{x^{10}}$

4.7 $y = (x^2 + 1)(2x + 3)$

4.8 $y = (x^2 + 1)(x^2 - 1)$

4.9 $y = \frac{5x^3 + 2x^2 - 3}{x}$

4.10 $y = \frac{x^2 + 1}{x^2 - 1}$

4.11 $y = \frac{1}{2 + x^3}$

4.12 $y = \frac{x^3}{x + 1}$

4.13 $y = \frac{x^3 + x}{2x^2 + 1}$

4.14 $y = \frac{2x + 1}{2x - 1}$

4.15 $y = 5\sqrt{x} + \frac{1}{x}$

4.16 $y = \sqrt{x}(3x - 2)$

4.17 $y = \frac{1}{\sqrt{x}}$

4.18 Show that formula (4.9) for $\frac{d}{dx}x^{\frac{1}{2}}$ and your answer to Problem 4.17 for $\frac{d}{dx}x^{-\frac{1}{2}}$ both agree with the rule for integer exponents: $\frac{d}{dx}x^n = nx^{n-1}$.

4.19 Show that $\frac{d}{dx}(u \cdot z \cdot w) = zw\frac{du}{dx} + uw\frac{dz}{dx} + uz\frac{dw}{dx}$.

In Problems 4.20 and 4.21 use the formula of Problem 4.19 to find $\frac{dy}{dx}$.

4.20 $y = (x + 1)(x^2 + 2)(x - 2)$

4.21 $y = x^{-3}(x + 3)(x^2 - 4)$

4.22 Check that $\frac{d}{dx}x^{-4} = -4x^{-5}$ by differentiating the product: $\frac{d}{dx}(x^{-2} \cdot x^{-2})$.

4.23 Check that $\frac{d}{dx}x^{-5} = -5x^{-6}$ by differentiating the product: $\frac{d}{dx}(x^{-2} \cdot x^{-3})$.

4.24 Find $\frac{d}{dx}x^{\frac{1}{3}}$. *Hint*: Use $(a - b)(a^2 + ab + b^2) = a^3 - b^3$. Multiply the top and bottom of $\frac{[(x+\Delta x)^{\frac{1}{3}} - x^{\frac{1}{3}}]}{\Delta x}$ by $a^2 + ab + b^2$ with $a = (x + \Delta x)^{\frac{1}{3}}$, $b = x^{\frac{1}{3}}$.

4.25 Use the product rule to show that if N is a positive integer and $\frac{d}{dx}x^N = Nx^{N-1}$, then $\frac{d}{dx}x^{N+1} = \frac{d}{dx}(x^N \cdot x) = (N + 1)x^N$. This is the essential step in the inductive proof that $\frac{d}{dx}x^n = nx^{n-1}$ for all positive integers.

5

The Chain Rule

Most of the functions we use are composites of simpler functions. For example, $(2x^2+1)^3$ is calculated by evaluating $2x^2+1$ and then cubing that number. If $f(x) = 2x^2+1$ and $g(x) = x^3$, then

$$g(f(x)) = f(x)^3 = (2x^2+1)^3.$$

Similarly, we can write $\sqrt{3x+1}$ as the composition of $\sqrt{x}$ and $3x+1$: if $f(x) = 3x+1$ and $g(x) = \sqrt{x}$, then

$$g(f(x)) = \sqrt{f(x)} = \sqrt{3x+1}.$$

Here are some other examples:

(i) $\dfrac{1}{x^2+4} = g(f(x))$ with $g(x) = \dfrac{1}{x}$, $f(x) = x^2+4$;

(ii) $(3x-1)^4 = g(f(x))$ with $g(x) = x^4$, $f(x) = 3x-1$;

(iii) $\dfrac{1}{\sqrt{x^2+x}} = h(g(f(x)))$ with $h(x) = \dfrac{1}{x}$, $g(x) = \sqrt{x}$, $f(x) = x^2+x$.

Using dependent variable notation (y, z, w, etc.) instead of functional notation ($f(x)$, $g(x)$, etc.), the above examples would look like this:

(i) if $z = \dfrac{1}{y}$ and $y = x^2+4$, then $z = \dfrac{1}{x^2+4}$;

(ii) if $z = y^4$ and $y = 3x-1$, then $z = (3x-1)^4$;

(iii) if $z = \frac{1}{w}$ and $w = \sqrt{y}$ and $y = x^2+x$, then $z = \frac{1}{\sqrt{x^2+x}}$.

Situations like these are quite common in physical applications. For example, if the force F on an object depends on its position x, and the position x depends on the time t, then F is a (composite) function of t. Because most functions that occur in practice are compositions of simpler functions, we proceed to find the rule for differentiating such functions.

Consider the general situation in which z is a function of y and y is a function of x, so that z also becomes a function of x. Let Δy be the change in y which results from changing x

by an amount Δx. Let Δz be the change in z which results from changing y by Δy, so Δz also corresponds to the change Δx in x. Notice that $\Delta y \longrightarrow 0$ as $\Delta x \longrightarrow 0$, since y is necessarily continuous if $\frac{dy}{dx}$ exists. Hence,

$$\begin{aligned} \frac{dz}{dx} &= \lim_{\Delta x \to 0} \frac{\Delta z}{\Delta x} \\ &= \lim_{\Delta x \to 0} \frac{\Delta z}{\Delta y} \cdot \frac{\Delta y}{\Delta x} \\ &= \lim_{\Delta x \to 0} \frac{dz}{dy} \cdot \frac{dy}{dx}. \end{aligned} \tag{5.1}$$

This is the important **chain rule** for differentiating composite functions.

In the argument leading to (5.1), we multiplied and divided by Δy, which is clearly not legitimate if $\Delta y = 0$ for arbitrarily small values of Δx. However, in that case Δz is also zero for arbitrarily small values of Δx, since z doesn't change if y doesn't change, and both $\frac{dy}{dx}$ and $\frac{dz}{dx}$ would then be zero, and (5.1) would still hold in the form $0 = 0$.

We illustrate the chain rule with the function $z = (2x^2+1)^3$, or $z = y^3$ with $y = 2x^2+1$. From (5.1),

$$\begin{aligned} \frac{dz}{dx} &= \frac{dz}{dy} \cdot \frac{dy}{dx} \\ &= 3y^2 \frac{dy}{dx} \\ &= 3y^2(4x) \\ &= 3(2x^2+1)^2(4x). \end{aligned}$$

In practice, we think of this computation as

$$\frac{d}{dx}(\quad)^3 = 3(\quad)^2 \frac{d}{dx}(\quad),$$

which holds no matter what function is inside the parentheses. For example, if $z = (\sqrt{x}+x^5)^3$, then

$$\begin{aligned} \frac{dz}{dx} &= 3(\quad)^2 \frac{d}{dx}(\quad) \\ &= 3\left(\sqrt{x}+x^5\right)^2 \cdot \frac{d}{dx}\left(\sqrt{x}+x^5\right) \\ &= 3\left(\sqrt{x}+x^5\right)^2 \cdot \left(\frac{1}{2\sqrt{x}} + 5x^4\right). \end{aligned}$$

To illustrate the chain rule with a different outside function, consider $y = 1/(3-x^2)^4$. Here

$$y = (\quad)^{-4},$$

so

$$\begin{aligned} \frac{dy}{dx} &= -4(\quad)^{-5} \frac{d}{dx}(\quad) \\ &= -4(3-x^2)^{-5} \frac{d}{dx}(3-x^2) \\ &= -4(3-x^2)^{-5}(-2x). \end{aligned}$$

Here are some more examples.

$$\frac{d}{dx}(3x-6x^3)^5 = 5(3x-6x^3)^4 \cdot (3-18x^2);$$

$$\frac{d}{dx}(5x^3+1)^{-2} = -2(5x^3+1)^{-3} \cdot (15x^2);$$

$$\frac{d}{dx}\frac{1}{1+x^2} = \frac{d}{dx}(1+x^2)^{-1} = -(1+x^2)^{-2}(2x).$$

Recall that $\frac{d}{dx}(\sqrt{x}) = \frac{1}{2\sqrt{x}}$, so the chain rule gives the general formula

$$\frac{d}{dx}\sqrt{(\ )} = \frac{1}{2\sqrt{(\ )}}\frac{d}{dx}(\ ).$$

For example,

$$\frac{d}{dx}\sqrt{3-5x} = \frac{1}{2\sqrt{3-5x}}(-5);$$

$$\frac{d}{dx}\sqrt{x^2+5x^4} = \frac{1}{2\sqrt{x^2+5x^4}}(2x+20x^3);$$

$$\frac{d}{dx}\sqrt{1+\frac{1}{x}} = \frac{1}{2\sqrt{1+\frac{1}{x}}}\left(-\frac{1}{x^2}\right).$$

Now consider $z = \frac{1}{\sqrt{1+x^2}}$. This is a composition of three functions:

$$z = \frac{1}{w},\ w = \sqrt{y},\ y = 1+x^2.$$

The derivative is

$$\begin{aligned}\frac{dz}{dx} &= \frac{dz}{dw}\cdot\frac{dw}{dy}\cdot\frac{dy}{dx}\\ &= -\frac{1}{w^2}\frac{1}{2\sqrt{y}}\cdot 2x\\ &= -\frac{1}{(1+x^2)}\cdot\frac{1}{2\sqrt{1+x^2}}\cdot 2x\\ &= -x/(1+x^2)^{\frac{3}{2}}.\end{aligned}$$

We could also find $\frac{dz}{dx}$ by the quotient rule in this example.

$$\begin{aligned}\frac{d}{dx}\frac{1}{\sqrt{1+x^2}} &= \frac{\sqrt{1+x^2}\frac{d}{dx}(1) - 1\cdot\frac{d}{dx}\sqrt{1+x^2}}{(1+x^2)}\\ &= \frac{-\frac{1}{2\sqrt{1+x^2}}(2x)}{(1+x^2)^2}\\ &= \frac{-x}{(1+x^2)^{\frac{3}{2}}}.\end{aligned}$$

EXAMPLE 5.1

Grain pours out of a spout at 5 cu ft/min onto a concrete floor and forms a conical pile whose radius and height are equal. At what rate is the height of the pile increasing when $h = 4$ ft?

Solution

The volume of a cone is $\frac{1}{3}\pi r^2 h$, and since here $r = h$, $V = \frac{1}{3}\pi h^3$. We are given that $\frac{dV}{dt} = 5$, and the chain rule shows that

$$\frac{dV}{dt} = \pi h^2 \frac{dh}{dt}.$$

Therefore, $5 = \pi h^2 \frac{dh}{dt}$, and $\frac{dh}{dt} = 5/\pi h^2$. When $h = 4$, $\frac{dh}{dt} = 5/16\pi$ ft/min.

EXAMPLE 5.2

A car drives east at 40 mph starting at town A, which is 60 miles due south of town B. At what rate is the distance (as the crow flies) between the car and town B increasing when $t = 2$ hr?

Solution

If s is the distance between the car and town B, and x is the distance from the car to town A, then

$$s^2 = 60^2 + x^2.$$

Hence,

$$2s\frac{ds}{dt} = 2x\frac{dx}{dt}.$$

When $t = 2$, $x = 80$ and

$$s = \sqrt{60^2 + 80^2} = \sqrt{10{,}000} = 100.$$

Since $\frac{dx}{dt} = 40$,

$$\frac{ds}{dt} = \frac{x}{s} \cdot 40 = \frac{80}{100} \cdot 40 = 32(\text{mi/hr}).$$

PROBLEMS

Find $\frac{dy}{dx}$.

5.1 $y = (1 + 3x)^4$

5.2 $y = (x + x^2)^5$

5.3 $y = (1 + 2x)^{-1}$

5.4 $y = (x^3 + x^7)^{-10}$

5.5 $y = \dfrac{1}{2 + 3x}$

5.6 $y = \dfrac{1}{1 + x^2}$

5.7 $y = \sqrt{1 + x^2}$

5.8 $y = (\sqrt{x} + 1)^3$

5.9 $y = \left(\dfrac{2x + 1}{1 + x^2}\right)^3$

5.10 $y = \left(\dfrac{\sqrt{x}}{x + 1}\right)^5$

5.11 $y = \dfrac{1}{\sqrt{x^3 - 2}}$

5.12 $y = \left(\sqrt{1 + x} + x\right)^{-2}$

Write $\frac{dz}{dx}$ in terms of x:

5.13 $z = y^{-3}, \quad y = 2x + 1$

5.14 $z = y^5, \quad y = \dfrac{1}{x + 1}$

5.15 $z = \frac{1}{y}, \quad y = \sqrt{x}$

5.16 $z = \sqrt{y}, \quad y = \frac{1}{x}$

5.17 $z = \frac{1}{w}, \quad w = \sqrt{y}, \quad y = x^2 + 9$

5.18 $z = w^2, \quad w = 1 + \sqrt{y}, \quad y = x + 1$

5.19 If a rectangle has sides x and y and x is increasing at 1 cm/sec and y at 3 cm/sec, how fast is the area increasing when $x = 4$ and $y = 5$? *Hint*:

$$\frac{dA}{dt} = \frac{d}{dt}(xy) = x\frac{dy}{dt} + y\frac{dx}{dt}.$$

5.20 In Example 5.1 suppose the grain flows out of the spout at 3 cu ft/min and the pile retains the shape of a cone with height equal to the diameter of the base. At what rate is the height increasing when $h = 6$? At what rate is the area of the base of the cone increasing when $h = 6$?

5.21 The temperature on a rod is given by $T = 3 + \sqrt{x}$ where x is the distance from the left end in inches and T is measured in degrees Centigrade. A sensor moves along the rod so that its position after t seconds is $x = t + \frac{1}{8}t^3$. At what rate is the temperature at the sensor changing when $t = 2$?

5.22 Water flows into a cylindrical bucket at 10 cu in/sec. The bucket has diameter 1 ft. How fast is the water rising in the bucket? *Hint*: The volume V of water in the bucket is $\pi 6^2 y$ in^3, where y is the height of the water in inches.

5.23 A spherical balloon is inflated so that its volume at time t is $\sqrt{1+t}$. At what rate is the radius r increasing when $r = 5$? ($V = \frac{4}{3}\pi r^3$)

5.24 Two ships leave the same dock at the same time. One goes east at 20 knots and the other north at 10 knots. At what rate (nautical miles per hour) is the distance between them changing when $t = 3$ hr? (1 knot is one nautical mile (1.15 miles) per hour.)

5.25 The top half of the circle $x^2 + y^2 = a^2$ has the equation $y = \sqrt{a^2 - x^2}$. Find the slope of the tangent at $(x, \sqrt{a^2 - x^2})$, and show that the tangent is perpendicular to the radius to the point; that is, show that the tangent has slope $-x/y$.

5.26 An object moves along the x-axis so that its position at time t is given by $x = (t^2 - 4t)^3$. The object moves to the right when $\frac{dx}{dt} > 0$ and to the left when $\frac{dx}{dt} < 0$. When does the object change direction?

5.27 Suppose $f'(x) = \frac{1}{x}$. If $g(x) = f(x^2 + x)$, what is $g'(x)$?

6

Trigonometric Functions

In trigonometry courses, the functions $\sin x$ and $\cos x$ have to do with the *angle* x. If x is an acute angle in a right triangle (Figure 6.1), then $\sin x$ is the opposite side over the hypotenuse and $\cos x$ is the adjacent side over the hypotenuse. If the hypotenuse has length 1, then the opposite side has length $\sin x$ and the adjacent side has length $\cos x$.

In calculus we more often think of $\sin x$ and $\cos x$ as functions of a *number* x. Suppose you start at the point $(1, 0)$ and go x units around the unit circle in the counterclockwise direction. The point you arrive at has coordinates $(\cos x, \sin x)$ as indicated in Figure 6.2. For calculus purposes this is the definition of $\cos x$ and $\sin x$. The angle at the origin which cuts off an arc of length x on the unit circle is an angle of x **radians**. If x is a negative number, then $\cos x$ and $\sin x$ are the coordinates of the point which is $|x|$ units around the circle in the clockwise direction. The picture shows that for all x,

$$\cos(-x) = \cos x; \quad \sin(-x) = -\sin x.$$

Since the unit circle has length 2π, the angle 2π radians corresponds to $360°$. From this we can easily calculate the equivalent radian measure for any number of degrees. For example, $30°$ corresponds to $\frac{30}{360} \cdot 2\pi = \frac{\pi}{6}$ radians, and $90°$ corresponds to $\frac{90}{360} \cdot 2\pi = \frac{\pi}{2}$ radians. Table 6.1 gives the values of $\cos x$ and $\sin x$ for common values of x corresponding to angles in the first quadrant.

For angles outside the first quadrant (radian measure outside $[0, \frac{\pi}{2}]$), we read the values of $\cos x$ and $\sin x$ from the appropriate diagram as indicated in Figure 6.3. From the definition it is clear that if you go 2π units around the circle from the point x units from $(1, 0)$ you get back to the same point $(\cos x, \sin x)$. That is, for any x,

$$\sin(x + 2\pi) = \sin x,$$
$$\cos(x + 2\pi) = \cos x.$$

The functions $\cos x$ and $\sin x$ are thus **periodic functions** with **period** 2π.

The remaining four trigonometric functions are defined as follows:

$$\tan x = \frac{\sin x}{\cos x}; \quad \cot x = \frac{\cos x}{\sin x};$$

$$\sec x = \frac{1}{\cos x}; \quad \csc x = \frac{1}{\sin x}.$$

We will deal almost entirely with $\cos x$, $\sin x$, $\tan x$, and $\sec x$.

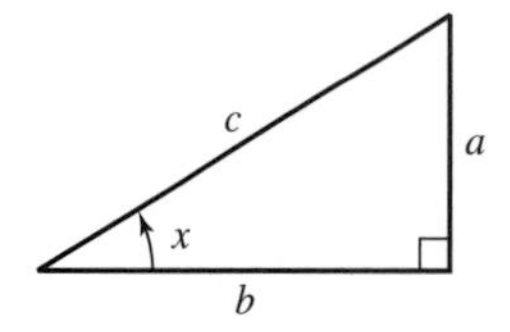

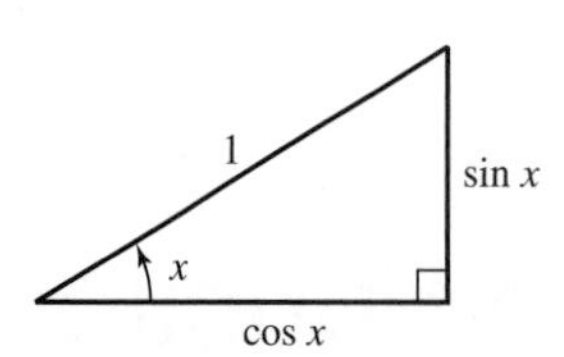

Figure 6.1

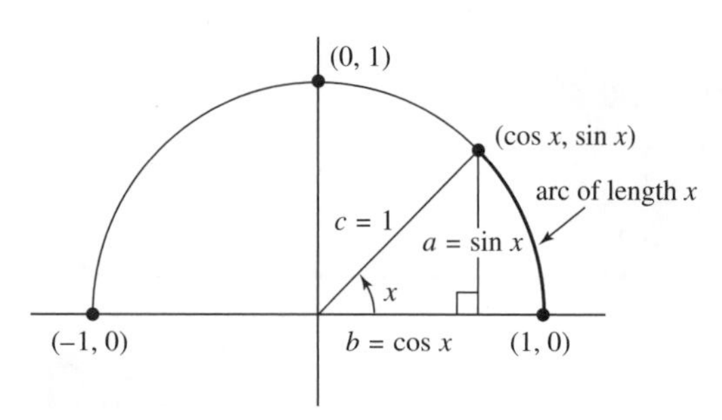

Figure 6.2

TABLE 6.1

Degrees	**0**	**30°**	**45°**	**60°**	**90°**
radians (x)	0	$\frac{\pi}{6}$	$\frac{\pi}{4}$	$\frac{\pi}{3}$	$\frac{\pi}{2}$
$\cos(x)$	1	$\frac{\sqrt{3}}{2}$	$\frac{\sqrt{2}}{2}$	$\frac{1}{2}$	0
$\sin(x)$	0	$\frac{1}{2}$	$\frac{\sqrt{2}}{2}$	$\frac{\sqrt{3}}{2}$	1

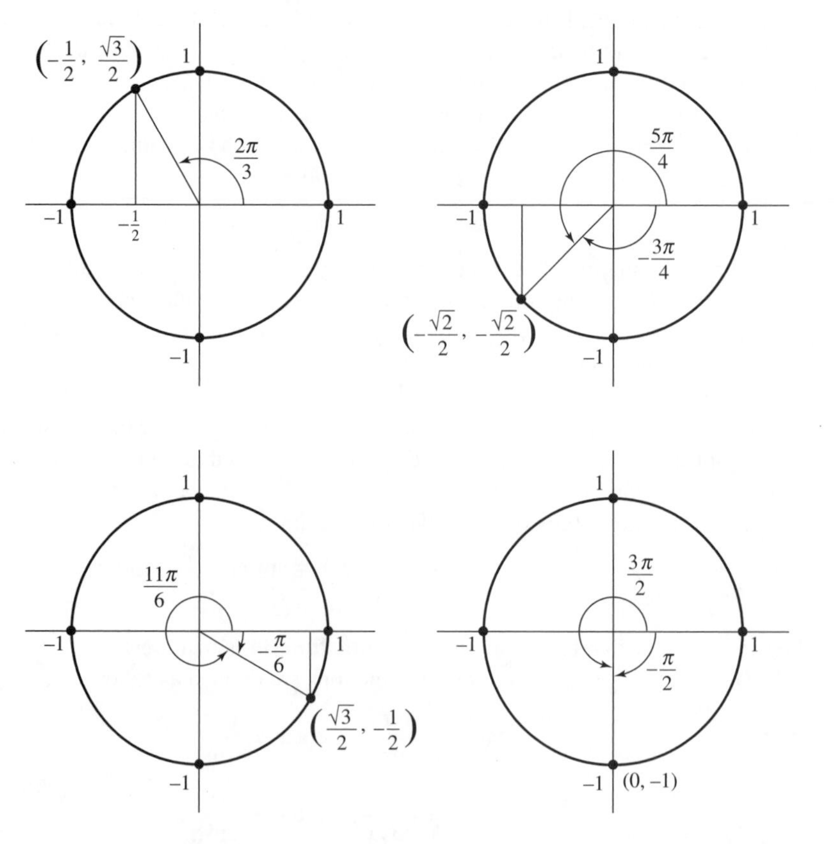

Figure 6.3

The following identities are used constantly in our calculations and should be memorized.

(i) $\sin^2 x + \cos^2 x = 1$

(ii) $\sin(x + y) = \sin x \cos y + \cos x \sin y$

$\sin(x - y) = \sin x \cos y - \cos x \sin y$

(iii) $\sin 2x = 2 \sin x \cos x$

(iv) $\cos(x + y) = \cos x \cos y - \sin x \sin y$

$\cos(x - y) = \cos x \cos y + \sin x \sin y$

(v) $\cos 2x = \cos^2 x - \sin^2 x$

$= 1 - 2\sin^2 x$

$= 2\cos^2 x - 1$

(vi) $\cos^2 x = \frac{1}{2}(1 + \cos 2x)$

(vii) $\sin^2 x = \frac{1}{2}(1 - \cos 2x)$

(viii) $\tan^2 x + 1 = \sec^2 x$.

The first identity, (i), simply says that $(\cos x, \sin x)$ is a point on the unit circle. Identities (ii) and (iv) are not too hard to verify, but the verification does nothing to uplift the spirit and so will be omitted. Identities (iii) and (v) for $\sin 2x$ and $\cos 2x$, respectively, follow from (ii) and (iv) by substituting x for y. The alternative forms for $\cos 2x$ in (v) result from replacing $\cos^2 x$ by $1 - \sin^2 x$ or $\sin^2 x$ by $1 - \cos^2 x$. These last two equations give the formulas (vi) and (vii) for $\cos^2 x$ and $\sin^2 x$ in terms of $\cos 2x$, and these will be used later to integrate $\cos^2 x$ and $\sin^2 x$. The final identity involving $\tan^2 x$ and $\sec^2 x$ results by dividing the identity $\sin^2 x + \cos^2 x = 1$ through by $\cos^2 x$.

To differentiate the trigonometric functions, we will need the following limits:

$$\lim_{x \to 0} \frac{\sin x}{x} = 1, \tag{6.1}$$

$$\lim_{x \to 0} \frac{1 - \cos x}{x} = 0. \tag{6.2}$$

The geometry of Figure 6.4 suggests that

$$\sin x < x < \frac{\sin x}{\cos x},$$

and these inequalities do indeed hold for all x with $0 < x < \frac{\pi}{2}$. From the first inequality we get

$$\frac{\sin x}{x} < 1,$$

and from the second we get

$$\cos x < \frac{\sin x}{x}.$$

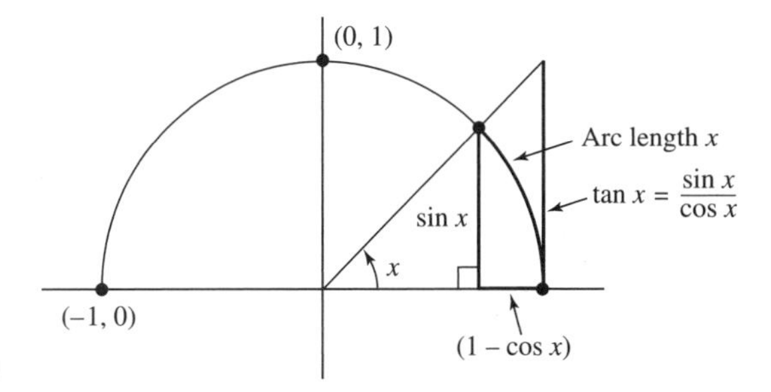

Figure 6.4

Putting these together we have

$$\cos x < \frac{\sin x}{x} < 1. \tag{6.3}$$

Since $\cos x \longrightarrow 1$ as $x \longrightarrow 0$, $\sin x/x \longrightarrow 1$. To see what happens as $x \longrightarrow 0$ through negative values, notice that $\cos(-x) = \cos x$ and $\sin(-x)/(-x) = \sin x/x$, so (6.3) still holds for negative x.

To verify (6.2) we notice from Figure 6.4 that the arc of length x is longer than the hypotenuse of the right triangle whose sides are $\sin x$ and $1 - \cos x$. Therefore,

$$\begin{aligned} \sin^2 x + (1 - \cos x)^2 &\le x^2, \\ \sin^2 x + \cos^2 x + 1 - 2\cos x &\le x^2, \\ 2 - 2\cos x &\le x^2, \\ \frac{1 - \cos x}{x} &\le \frac{1}{2}x. \end{aligned}$$

Hence, $(1 - \cos x)/x \longrightarrow 0$ as $x \longrightarrow 0$, which is (6.2). This limit can also be deduced from (6.1) by algebraic manipulation as follows:

$$\begin{aligned} \frac{1 - \cos x}{x} &= \frac{(1 - \cos x)(1 + \cos x)}{x(1 + \cos x)} \\ &= \frac{1 - \cos^2 x}{x(1 + \cos x)} = \frac{\sin x}{x}\,\frac{\sin x}{1 + \cos x}. \end{aligned}$$

Since $\frac{\sin x}{x} \longrightarrow 1$ and $\frac{\sin x}{(1+\cos x)} \longrightarrow 0$ as $x \longrightarrow 0$, we again get that $\frac{1-\cos x}{x} \longrightarrow 0$.

Now we can calculate the derivatives of $\sin x$ and $\cos x$.

$$\begin{aligned} \frac{d}{dx}\sin x &= \lim_{\Delta x \to 0} \frac{\sin(x + \Delta x) - \sin x}{\Delta x} \\ &= \lim_{\Delta x \to 0} \frac{\sin x \cos \Delta x + \cos x \sin \Delta x - \sin x}{\Delta x} \\ &= \lim_{\Delta x \to 0} \sin x \frac{(\cos \Delta x - 1)}{\Delta x} + \cos x \frac{\sin \Delta x}{\Delta x} \\ &= (\sin x) \cdot 0 + (\cos x) \cdot 1 \\ &= \cos x. \end{aligned}$$

We differentiate $\cos x$ using the identity

$$\begin{aligned} \sin\left(x + \frac{\pi}{2}\right) &= \sin x \cos \frac{\pi}{2} + \cos x \sin \frac{\pi}{2} \\ &= (\sin x) \cdot 0 + (\cos x) \cdot 1 = \cos x. \end{aligned}$$

Therefore,

$$\begin{aligned} \frac{d}{dx}\cos x &= \frac{d}{dx}\sin\left(x + \frac{\pi}{2}\right) \\ &= \cos\left(x + \frac{\pi}{2}\right) \\ &= \cos x \cos \frac{\pi}{2} - \sin x \sin \frac{\pi}{2} \\ &= (\cos x) \cdot 0 - (\sin x) \cdot 1 \\ &= -\sin x. \end{aligned}$$

The derivatives of $\tan x$ and $\sec x$ are easily calculated (Problem 6.2), and the four basic trigonometric differentiation formulas are

$$\frac{d}{dx}\sin x = \cos x; \quad \frac{d}{dx}\cos x = -\sin x;$$

$$\frac{d}{dx}\tan x = \sec^2 x; \quad \frac{d}{dx}\sec x = \sec x \tan x.$$

EXAMPLE 6.1

Find $\frac{dy}{dx}$ if

$$\text{(i)}\ y = \sin(x^3 + 1);$$
$$\text{(ii)}\ y = \cos 3x + \tan^2 x;$$
$$\text{(iii)}\ y = \sec \sqrt{x}.$$

Solution

Each of these problems involves a chain rule differentiation.

$$\text{(i)}\ \frac{dy}{dx} = \frac{d}{dx}\sin(x^3 + 1) = [\cos(x^3 + 1)]3x^2;$$

$$\text{(ii)}\ \frac{dy}{dx} = \frac{d}{dx}(\cos 3x + \tan^2 x) = -3\sin 3x + 2\tan x \sec^2 x;$$

$$\text{(iii)}\ \frac{dy}{dx} = \frac{d}{dx}(\sec\sqrt{x}) = (\sec\sqrt{x}\tan\sqrt{x})\frac{1}{2\sqrt{x}}.$$

EXAMPLE 6.2

A searchlight, rotating at 3 radians/min, plays a spot on a wall that is 100 ft away. At what rate is the lighted spot traveling along the wall when the beam is perpendicular to the wall?

Solution

Measure distance along the wall from the point, 0, directly opposite the light (Figure 6.5). We are given that $\frac{d\theta}{dt} = 3$. Since $x/100 = \tan\theta$,

$$\frac{dx}{dt} = \frac{d}{dt}100\tan\theta$$

$$= 100\sec^2\theta\frac{d\theta}{dt}.$$

When $x = 0$, $\theta = 0$, $\sec^2\theta = 1$, then $\frac{dx}{dt} = 300$ ft/min or 5 ft/sec.

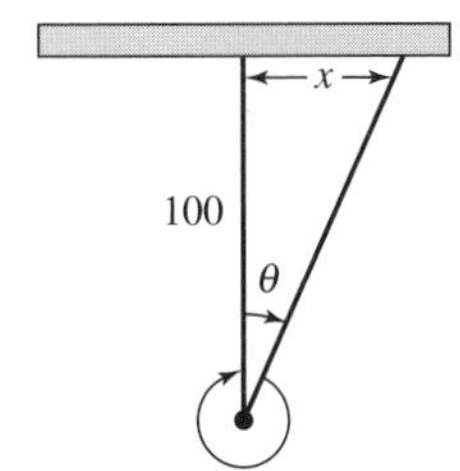

Figure 6.5

PROBLEMS

6.1 Graph $y = \sin x$ and $y = \cos x$ on the same coordinate system for $0 \le x \le 2\pi$. Use your calculator (set for radians, not degrees) to find values for $x = 0, 0.5, 1.0, 1.5, \ldots, 5.5, 6.0, 6.28$.

6.2 Derive the formulas $\frac{d}{dx}\tan x = \sec^2 x$ and $\frac{d}{dx}\sec x = \sec x \tan x$, using $\frac{d}{dx}\sin x = \cos x$ and $\frac{d}{dx}\cos x = -\sin x$.

Find $\frac{dy}{dx}$.

6.3 $y = \cos 2x$

6.4 $y = \sin(3x + 1)$

6.5 $y = \cos^2 x$

6.6 $y = \cos x \sin x$

6.7 $y = \sin^2 8x$

6.8 $y = \cos^3(5x + 1)$

6.9 $y = \sqrt{1 + \cos x}$

6.10 $y = (\sin x + \cos x)^2$

6.11 $y = x \sec x$

6.12 $y = \tan(4x)$

6.13 $y = \sec x \tan x$

6.14 $y = \sec^2 x - 1$

6.15 $y = \tan^2 x$

6.16 $y = \sec^3 x$

6.17 $y = \dfrac{\sin x}{1 + \cos x}$

6.18 $y = (\tan x + 1)^2$

6.19 $y = (2x^2 + 1)\cos 3x$

6.20 $y = \dfrac{\sin 5x}{x}$

6.21 $y = \dfrac{x^2 + 1}{\sin x}$

6.22 $y = \dfrac{x \sin x}{\cos x}$

6.23 Derive identities (vi) and (vii) from (v).

6.24 Show that $\frac{1}{2}\sin x = \sin\frac{x}{2}\cos\frac{x}{2}$.

6.25 Show that $\tan(x + y) = \frac{\tan x + \tan y}{1 - \tan x \tan y}$.

6.26 In the situation of Example 6.2, at what rate is the lighted spot traveling when the beam makes an angle of 45° with the wall?

6.27 A pendulum swings so that the angle θ the rod makes with the vertical is given by $\theta = \frac{\pi}{12}\sin(\omega t)$, where ω is a constant and t is measured in seconds.

(i) If the pendulum is at its maximum swing from the vertical for the first time at $t = 1$, what is ω, and what is the next time the pendulum is at its maximum swing on the other side of its arc?

(ii) Let x be the point on the ground directly under the pendulum, with $x = 0$ when $t = 0$. If the pendulum is 30 in long, what is $\frac{dx}{dt}$ when $t = 1$? When $t = 2$? $t = 3$?

6.28 Let θ be the angle a ladder makes with a wall. If the ladder is 10 ft long and the base is pulled away from the wall at 3 ft/sec, what is $\frac{d\theta}{dt}$ when the top of the ladder is 5 ft above the floor?

6.29 A boat passes point A going east at 44 ft/sec. A searchlight at B, 200 feet south of point A, follows the boat. At what rate is the searchlight turning after 3 seconds? *Hint*: Let θ be the angle between the line AB and the line from B to the boat; find $\frac{d\theta}{dt}$.

6.30 Show that $\cos x = \sin(x + \frac{\pi}{2})$.

6.31 Show that for any two numbers a and b such that $a^2 + b^2 = 1$ there is a number θ such that $a = \cos\theta$, $b = \sin\theta$. How many such numbers θ are there?

6.32 Every curve $y = A\sin x + B\cos x$ is a sine curve; that is, for all A and B there are numbers K and θ such that $A\sin x + B\cos x = K\sin(x + \theta)$. *Hint*: Write

$$A\sin x + B\cos x = \sqrt{A^2 + B^2}\left[(\sin x)\frac{A}{\sqrt{A^2 + B^2}} + (\cos x)\frac{B}{\sqrt{A^2 + B^2}}\right].$$

Notice that if $a = \frac{A}{\sqrt{A^2+B^2}}$ and $b = \frac{B}{\sqrt{A^2+B^2}}$, then $a^2 + b^2 = 1$. Consult Problem 6.31.

Find all points x in $[0, 2\pi]$ where the tangent line is horizontal.

6.33 $y = \sin x$

6.34 $y = \cos x$

6.35 $y = \sin x + \cos x$

6.36 $y = \sin^2 x$

6.37 $y = x - \tan x$

7

Exponential Functions and Logarithms

The **exponential functions** are the functions of the form $y = a^x$, where a is a positive constant. For example,

$$y = 2^x, \quad y = \left(\frac{1}{2}\right)^x, \quad y = 1.06^x, \quad y = 10^x$$

are exponential functions.

Consider for a moment what a^x means for a positive number a and an arbitrary real number x. If x is an integer, there is no problem: a^2, a^5, a^{-3}—these expressions are part of arithmetic. If the exponent x is a fraction, then a^x is no longer just multiplication or division. For example, $a^{\frac{1}{2}}$, $a^{\frac{1}{3}}$ are the roots of a, and it is a fact that we here take for granted that a positive number has a square root, cube root, and so on. If x is a fraction, say $x = \frac{p}{q}$ for positive integers p and q, then $a^{\frac{p}{q}}$ is the pth power of the qth root of a. That takes care of rational exponents x. If x is an irrational number like $\sqrt{2}$ or π, then a^x is the limit of the numbers $a^{\frac{p}{q}}$ where the rational numbers $\frac{p}{q}$ approach x. It is a theorem that the limit exists and that exponentials defined this way satisfy all the usual rules for all exponents x.

Notice the distinction between the exponential function a^x and the power function x^n. In the power functions we have considered so far, $y = x^2$, $y = x^{\frac{1}{2}}$, $y = x^{-3}$, and so on, the base x is the variable, and the exponent is fixed. In the exponential function a^x, the exponent is the variable. The graphs of $y = 2^x$ and $y = (\frac{1}{2})^x$ are shown in Figure 7.1. Notice that the two curves are symmetric about the y-axis, since $(\frac{1}{2})^x = 2^{-x}$.

The curve $y = a^x$ for $a > 1$ will rise steeply as x increases, and decrease rapidly to zero as $x \longrightarrow -\infty$. If $0 < a < 1$, then the curve decreases rapidly to zero as $x \longrightarrow \infty$, and increases sharply as $x \longrightarrow -\infty$.

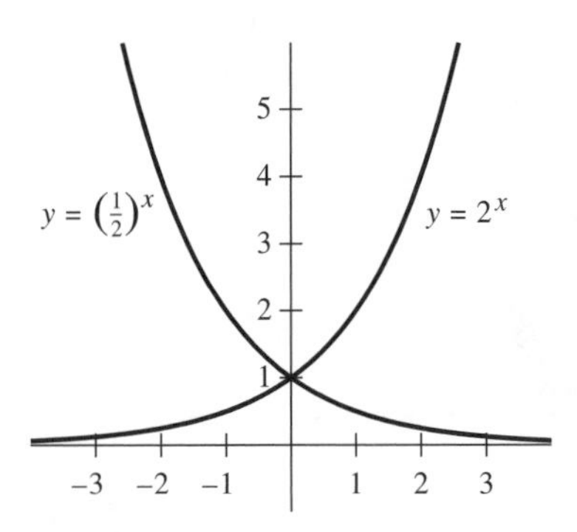

Figure 7.1

We calculate $\frac{d}{dx}a^x$ as follows:

$$\begin{aligned}\frac{d}{dx}a^x &= \lim_{\Delta x \to 0} \frac{a^{x+\Delta x} - a^x}{\Delta x} \\ &= \lim_{\Delta x \to 0} \frac{a^x a^{\Delta x} - a^x}{\Delta x} \\ &= \lim_{\Delta x \to 0} a^x \frac{(a^{\Delta x} - 1)}{\Delta x} \\ &= a^x \lim_{\Delta x \to 0} \frac{a^{\Delta x} - 1}{\Delta x}.\end{aligned}$$

It can be shown that this limit exists for any $a > 0$, and we let

$$k_a = \lim_{\Delta x \to 0} \frac{a^{\Delta x} - 1}{\Delta x}.$$

With this notation we can write

$$\frac{d}{dx}a^x = k_a a^x. \tag{7.1}$$

The derivative of any exponential function (i.e., any a) is always proportional to the function value.

Now we do some experimenting to see how the proportionality constant k_a depends on a. First let $a = 2$ and calculate $(2^{\Delta x} - 1)/\Delta x$ for small values of Δx.

Δx	0.1	.01	.001	.0001
$2^{\Delta x} - 1$	.0718	.00695	.000693	.000693
$(2^{\Delta x} - 1)/\Delta x$	.718	.695	.693	.693

As far as our eight-place calculator can tell

$$k_2 = \lim_{\Delta x \to 0} \frac{2^{\Delta x} - 1}{\Delta x} \doteq .693.$$

If we do the same calculations (Problem 7.1) for $a = 3$, we get the approximate limit

$$k_3 = \lim_{\Delta x \to 0} \frac{3^{\Delta x} - 1}{\Delta x} \doteq 1.099.$$

It is clear that the difference quotient $(a^{\Delta x} - 1)/\Delta x$ gets bigger as a increases, for any fixed Δx. Therefore, k_a increases as a increases. It is therefore plausible that there is some number

e between 2 and 3 such that $k_e = 1$, and experimentation shows (Problem 7.2) that e is approximately equal to 2.718. Now for this number e,

$$\frac{d}{dx}e^x = e^x, \tag{7.2}$$

and the elegance of (7.2) is the justification for introducing the number e. The chain rule then gives the formulas

$$\frac{d}{dx}e^{f(x)} = e^{f(x)} f'(x),$$

or

$$\frac{d}{dx}e^u = e^u \frac{du}{dx}.$$

As examples, we have

$$\frac{d}{dx}e^{x^2} = e^{x^2} \cdot 2x;$$

$$\frac{d}{dx}e^{-3x} = e^{-3x}(-3);$$

$$\frac{d}{dx}e^{\sqrt{x}} = e^{\sqrt{x}}\frac{1}{2\sqrt{x}};$$

$$\frac{d}{dx}e^{\sin x} = e^{\sin x}\cos x.$$

The graph of $y = e^x$ is shown in Figure 7.2, along with its tangent line at (0, 1).

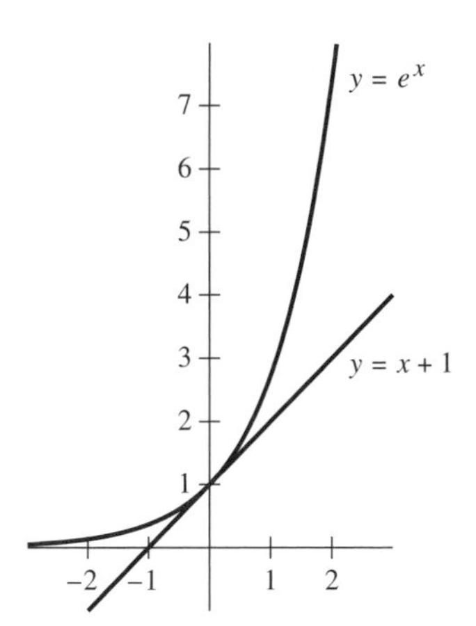

Figure 7.2

The function e^x is strictly increasing, is not bounded above, and approaches zero as $x \longrightarrow -\infty$. Every positive number y is e^x for some unique number x. This number x is called the natural logarithm of y and is denoted $\log y$. In other words, $y = e^x$ and $x = \log y$ mean the same thing. The functions e^x and $\log x$ are **inverse functions**; e^x is defined for all x and takes as values all positive numbers; $\log x$ is defined for all positive numbers and takes as values all numbers. For all $x > 0$, $e^{\log x} = x$, and for all x, $\log e^x = x$. The graphs of $y = e^x$ and $y = \log x$ are shown together in Figure 7.3.

Notice that the graphs are symmetric about the line $y = x$. This is true of every pair of inverse functions and simply reflects the fact that (a, b) is on the graph of one function if and only if (b, a) is on the graph of its inverse.

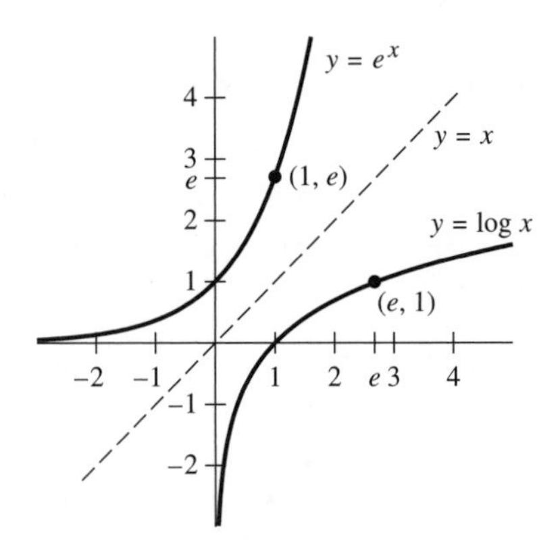

Figure 7.3

In general, the notation $f^{-1}(x)$ is used for the function inverse to $f(x)$, so $f\big(f^{-1}(x)\big) = f^{-1}(f(x)) = x$. Notice that the exponent notation is not used consistently here: $f(x)^2$ means $f(x) \cdot f(x)$, but $f^{-1}(x)$ never means $1/f(x)$.

The logarithm of a number is an exponent; that is, $\ell = \log N$ means $e^\ell = N$. Therefore, the rules for logarithms are just the algebraic rules for exponents. The log of a product is the sum of the logs since $e^{\ell_1} \cdot e^{\ell_2} = e^{\ell_1 + \ell_2}$. That is,

$$\log(ab) = \log a + \log b.$$

Similarly,

$$\log N^k = k \log N$$

since if $e^\ell = N$, $e^{k\ell} = N^k$, and the log of N^k is k times the log of N.

To find the derivative of $\log x$, we let $y = \log x$, so $e^y = x$, and then differentiate:

$$\begin{aligned} e^y &= x, \\ e^y \frac{dy}{dx} &= 1, \\ \frac{dy}{dx} &= \frac{1}{e^y} = \frac{1}{x}. \end{aligned}$$

The general chain rule formula is

$$\frac{d}{dx} \log u = \frac{1}{u} \frac{du}{dx}.$$

For example,

$$\begin{aligned} \frac{d}{dx} \log(x^2 + 1) &= \frac{2x}{x^2 + 1}; \\ \frac{d}{dx} \log \cos x &= -\frac{\sin x}{\cos x}; \\ \frac{d}{dx} \log(1 + e^x) &= \frac{e^x}{1 + e^x}. \end{aligned}$$

Many times a logarithm formula can be simplified before it is differentiated; for example,

$$\begin{aligned}\frac{d}{dx}\log[(x+1)(x^3-3)] &= \frac{d}{dx}[\log(x+1)+\log(x^3-3)]\\ &= \frac{1}{x+1}+\frac{3x^2}{x^3-3};\\ \frac{d}{dx}\log\frac{1}{(5x+1)^2} &= \frac{d}{dx}\log(5x+1)^{-2}\\ &= \frac{d}{dx}(-2)\log(5x+1)\\ &= (-2)\cdot\frac{5}{5x+1};\\ \frac{d}{dx}\log 2^x &= \frac{d}{dx}x\log 2 = \log 2.\end{aligned}$$

We chose e as the primary base for an exponential function because $k_e = 1$; that is, because $\frac{d}{dx}e^x = e^x \cdot 1$. Now let's go back to $\frac{d}{dx}a^x$ for general a. Since $\log a^x = x\log a$ and $u = e^{\log u}$ for all u, we can write

$$a^x = e^{\log a^x} = e^{x\log a}.$$

From this we get

$$\begin{aligned}\frac{d}{dx}a^x &= e^{x\log a}\frac{d}{dx}(x\log a)\\ &= a^x\log a.\end{aligned}$$

Hence, the constant $k_a = \log a$. Earlier we computed the approximate value $k_2 \doteq .693$, so from $\frac{d}{dx}2^x = k_2 2^x$ and $\frac{d}{dx}2^x = 2^x\log 2$ we conclude that $\log 2 \doteq .693$.

Functions that involve the variable in both the base and the exponent do not fall under either of the rules

$$\frac{d}{dx}x^n = nx^{n-1};\ \frac{d}{dx}a^x = a^x\log a.$$

To differentiate such a function, we first take its logarithm. For example, let

$$y = (x^2+2)^{3x}.$$

Then

$$\begin{aligned}\log y &= 3x\log(x^2+2),\\ \frac{1}{y}\frac{dy}{dx} &= 3\log(x^2+2)+3x\cdot\frac{2x}{x^2+2}\\ \frac{dy}{dx} &= y\left[3\log(x^2+2)+\frac{6x^2}{x^2+2}\right]\\ &= (x^2+2)^{3x}\left[3\log(x^2+2)+\frac{6x^2}{x^2+2}\right].\end{aligned}$$

PROBLEMS

7.1 Approximate k_3 by evaluating $(3^{\Delta x}-1)/\Delta x$ for $\Delta x = .1, .01, .001, .0001$.

7.2 Approximate k_e by calculating $[(2.718)^{\Delta x}-1]/\Delta x$ for $\Delta x = .1, .01, .001, .0001$.

Find $\frac{dy}{dx}$.

7.3 $y = e^{2x}$

7.4 $y = e^{x^2+x}$

7.5 $y = 1/e^{x^2}$

7.6 $y = (e^{\sqrt{x}} + 1)^2$

7.7 $y = \sqrt{xe^x}$

7.8 $y = 3^x$

7.9 $y = 2^{x^2}$

7.10 $y = x^2e^{3x} + x\cos x$

7.11 $y = e^x \cos x + e^{-x} \sin x$

7.12 $y = e^x \log x$

7.13 $y = e^{x^2} \log(1 + x^3)$

7.14 $y = \log 3x$

7.15 $y = \log(x^2 + x)$

7.16 $y = \log[xe^x]$

7.17 $y = \log[(x + 1)(x + 2)]$

7.18 $y = \log(x^2/3^x)$

7.19 $y = x^x$

Hint: Find $\frac{d}{dx} \log y$ first.

7.20 $y = (2x + 1)^x$

7.21 $y = (\sin x + \cos x)^x$

7.22 The number y of bacteria at time t in any given colony is given by $y = Ae^{kt}$ for some constants A and k. If $y = 1000$ at $t = 0$ and $y = 3000$ at $t = 2$, what are A and k? What is y when $t = 4$?

7.23 Show that if $y = Ae^{kt}$ and $y = 2A$ when $t = t_0$, then y will double in any time interval t_0; that is, for all t,

$$Ae^{k(t+t_0)}/Ae^{kt} = 2.$$

7.24 If y is the number of bacteria in a culture, and $y = 500$ at $t = 0$, so $y = 500e^{kt}$ (t is hours), and y doubles every hour, what is k?

7.25 If an amount A of money is invested at 6%, compounded annually, then after n years the investment is worth $A(1.06)^n$. How long does it take to double your money at 6%?

7.26 A common rule of thumb says that money doubles at interest rate r% after n years, where $n = \frac{70}{r}$. For example, at 7% money doubles in 10 years. Explain, keeping in mind that $\log 2 \doteq .70$ and for small x (e.g., .05, .07, etc.), $\log(1 + x) \doteq x$.

7.27 Show that the only functions y such that $\frac{dy}{dx} = y$ are the functions $y = Ae^x$. *Hint*: Assume y is such a function; let $u = ye^{-x}$ and show that $\frac{du}{dx} \equiv 0$.

7.28 The **hyperbolic cosine** and **hyperbolic sine** are defined as follows:

$$\cosh x = \frac{e^x + e^{-x}}{2}, \quad \sinh x = \frac{e^x - e^{-x}}{2}.$$

Show that
(a) $\frac{d}{dx} \cosh x = \sinh x$ and $\frac{d}{dx} \sinh x = \cosh x$;
(b) $\cosh^2 x - \sinh^2 x = 1$.

7.29 (a) Graph the curve $y = e^{\frac{1}{2}x}$, $-1 \le x \le 3$.
(b) Find the point $P = (x, e^{ax})$ on the curve $y = e^{ax}$ such that the tangent line at P passes through the origin. Does your answer look right if $a = \frac{1}{2}$?
(c) Find the point where the line perpendicular to $y = e^{ax}$ at P intersects the x-axis.

7.30 (a) Find the point(s) on the graph of $y = xe^x$ where the tangent line is horizontal.
(b) Graph the curve for $-5 \le x \le 1$.

8

Inverse Functions

A function f is said to be **one-to-one** on an interval I provided f does not take the same value for different values of x in I. For the continuous functions we consider here, a one-to-one function is either strictly increasing on I ($f(x_2) > f(x_1)$ if $x_2 > x_1$), or strictly decreasing ($f(x_2) < f(x_1)$ if $x_2 > x_1$). If f is a one-to-one function, then for each function value y there is a unique x such that $f(x) = y$. This mapping – from y to x – is denoted f^{-1} (read f-inverse), so that $x = f^{-1}(y)$ and $y = f(x)$ mean the same thing. In the last chapter we saw that $\log x$ and e^x are inverse functions, so that if $f(x) = e^x$, $f^{-1}(x) = \log x$. For any one-to-one function f, $f(f^{-1}(x)) = x$ and $f^{-1}(f(x)) = x$.

Now consider the inverses of the power functions x^n. If $n = 1, 3, 5, \ldots$, then x^n is strictly increasing for all x, and if $n = 2, 4, 6, \ldots$, then x^n is strictly increasing for $x \geq 0$. We use $x^{\frac{1}{n}}$ to denote the inverse of x^n, so if $f(x) = x^n$, then $f^{-1}(x) = x^{\frac{1}{n}}$, and $x^{\frac{1}{n}}$ is the nth root of x. If n is odd, then $x^{\frac{1}{n}}$ is defined for all x, but if n is even, $x^{\frac{1}{n}}$ is defined only for $x \geq 0$. Thus, $x^{\frac{1}{2}}$ denotes the positive square root of the positive number x.

We let $y = x^{\frac{1}{n}}$ and calculate $\frac{dy}{dx}$. Since $y^n = x$, we differentiate and get

$$ny^{n-1}\frac{dy}{dx} = 1,$$

$$\frac{dy}{dx} = \frac{1}{ny^{n-1}}.$$

Now write y^{n-1} in terms of x. Since $y = x^{\frac{1}{n}}$,

$$\begin{aligned}\frac{dy}{dx} &= \frac{1}{n\left(x^{\frac{1}{n}}\right)^{n-1}}\\ &= \frac{1}{n}\,\frac{1}{x^{1-\frac{1}{n}}}\\ &= \frac{1}{n}\,x^{\frac{1}{n}-1}.\end{aligned}$$

The rule for differentiating $x^{\frac{1}{n}}$ is the same as that for differentiating x^n or x^{-n}: multiply by the exponent, and subtract one from the exponent to get the new exponent. The chain rule extends the exponent rule to fractional exponents ($x^{\frac{m}{n}} = (x^{\frac{1}{n}})^m$) and negative exponents ($x^{-\frac{m}{n}} = (x^{\frac{m}{n}})^{-1}$):

$$\frac{d}{dx}x^{\frac{m}{n}} = \frac{m}{n}x^{\frac{m}{n}-1}; \quad \frac{d}{dx}x^{-\frac{m}{n}} = -\frac{m}{n}x^{-\frac{m}{n}-1}.$$

Indeed, the exponent rule holds for *any* exponent:

$$\frac{d}{dx}x^r = rx^{r-1}.$$

Here are some examples:

$$\frac{d}{dx}x^{\frac{1}{3}} = \frac{1}{3}x^{-\frac{2}{3}};$$
$$\frac{d}{dx}x^{\frac{4}{5}} = \frac{4}{5}x^{-\frac{1}{5}};$$
$$\frac{d}{dx}x^{\frac{7}{3}} = \frac{7}{3}x^{\frac{4}{3}};$$
$$\frac{d}{dx}x^{-\frac{5}{2}} = -\frac{5}{2}x^{-\frac{7}{2}};$$
$$\frac{d}{dx}x^{\sqrt{2}} = \sqrt{2}x^{\sqrt{2}-1}.$$

Now consider the inverses of the trigonometric functions. The functions $\sin x$ and $\cos x$ are periodic ($\sin x = \sin(x + 2\pi)$) and so are not one-to-one. However, the functions are one-to-one on certain intervals. For example, $\sin x$ is one-to-one (strictly increasing) on $[-\frac{\pi}{2}, \frac{\pi}{2}]$, and $\cos x$ is one-to-one (strictly decreasing) on $[0, \pi]$. We define $\sin^{-1} x$ and $\cos^{-1} x$ to be the inverses of the functions on these restricted domains. This is just what we did when we defined $x^{\frac{1}{2}}$ to be the inverse of x^2 restricted to $[0, \infty)$. So $\sin^{-1} x$, for $-1 \leq x \leq 1$, is the unique y between $-\frac{\pi}{2}$ and $\frac{\pi}{2}$ such that $\sin y = x$. Similarly, $\cos^{-1} x$ is the unique number y between 0 and π such that $\sin y = x$.

Now we calculate the derivatives. Let $y = \sin^{-1} x$, so

$$\begin{aligned} \sin y &= x, \\ \cos y \frac{dy}{dx} &= 1, \\ \frac{dy}{dx} &= \frac{1}{\cos y}, \\ &= \frac{1}{\cos(\sin^{-1} x)}. \end{aligned} \tag{8.1}$$

A similar process shows that if $y = \cos^{-1} x$, then

$$\frac{dy}{dx} = \frac{d}{dx}\cos^{-1} x = -\frac{1}{\sin(\cos^{-1} x)}. \tag{8.2}$$

To simplify the expressions $\cos(\sin^{-1} x)$ and $\sin(\cos^{-1} x)$, we draw the right triangle (Figure 8.1) in which the angles $\theta = \sin^{-1} x$ and $\varphi = \cos^{-1} x$ are displayed. From the figure we can read the identities

$$\begin{aligned} \cos(\sin^{-1} x) &= \cos\theta = \sqrt{1 - x^2}, \\ \sin(\cos^{-1} x) &= \sin\varphi = \sqrt{1 - x^2}. \end{aligned} \tag{8.3}$$

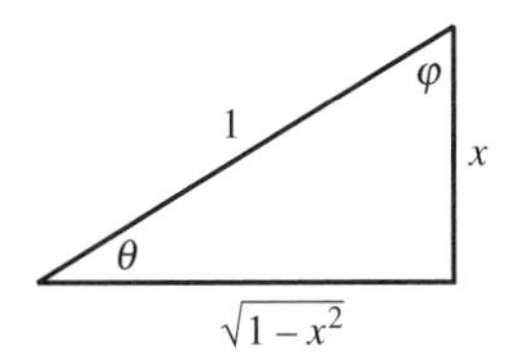

Figure 8.1

Now rewriting (8.1) and (8.2) using (8.3), we have

$$\frac{d}{dx}\sin^{-1}x = \frac{1}{\sqrt{1-x^2}},$$
$$\frac{d}{dx}\cos^{-1}x = -\frac{1}{\sqrt{1-x^2}}.$$

Here are some chain rule examples with $\sin^{-1}x$ and $\cos^{-1}x$:

$$\frac{d}{dx}\sin^{-1}(3x) = \frac{3}{\sqrt{1-(3x)^2}};$$
$$\frac{d}{dx}\cos^{-1}x^2 = -\frac{1}{\sqrt{1-x^4}}\cdot(2x);$$
$$\frac{d}{dx}\sin^{-1}e^x = \frac{e^x}{\sqrt{1-e^{2x}}};$$
$$\frac{d}{dx}\cos^{-1}(\log x) = -\frac{1}{\sqrt{1-(\log x)^2}}\frac{1}{x}.$$

Notice that all these functions have limited domains:

$$\sin^{-1}3x \text{ is defined if } -1 \le 3x \le 1;$$
$$\cos^{-1}x^2 \text{ is defined if } -1 \le x \le 1;$$
$$\sin^{-1}e^x \text{ is defined if } -\infty < x \le 0.$$

The function $\tan x$ is strictly increasing for $-\frac{\pi}{2} < x < \frac{\pi}{2}$ and takes on all real values on this interval. Therefore, $\tan^{-1}x$ is defined for all x and takes values in $(-\frac{\pi}{2}, \frac{\pi}{2})$. To find $\frac{d}{dx}\tan^{-1}x$ we let $y = \tan^{-1}x$, so

$$\tan y = x,$$
$$\sec^2 y\frac{dy}{dx} = 1,$$
$$\frac{dy}{dx} = \cos^2 y.$$

From Figure 8.2 we see that $\cos y = \frac{1}{\sqrt{1+x^2}}$, so

$$\frac{d}{dx}\tan^{-1}x = \frac{1}{1+x^2}. \tag{8.4}$$

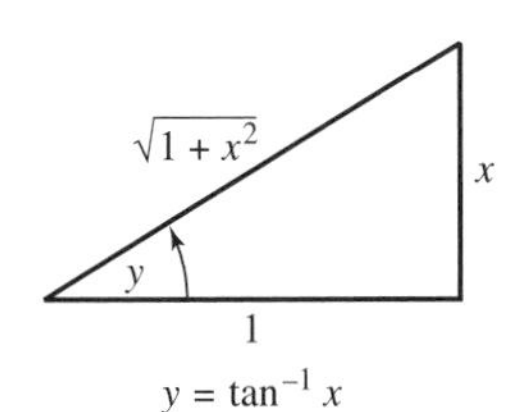

Figure 8.2

A variant of (8.4) that is important for our later work in integration is

$$\frac{d}{dx}\frac{1}{a}\tan^{-1}\frac{x}{a} = \frac{1}{a^2+x^2}. \tag{8.5}$$

PROBLEMS

Find $\frac{dy}{dx}$:

8.1 $y = 1 + x^{\frac{2}{3}}$
8.2 $y = x^{-\frac{3}{4}} + 5x^{\frac{1}{4}}$
8.3 $y = 3x^{\frac{7}{5}} - 2x^{-\frac{7}{5}}$
8.4 $y = x\sqrt{x} - 3x^2\sqrt{x}$
8.5 $y = (x^2+1)^{-\frac{2}{3}}$
8.6 $y = \sin(x^{\frac{2}{3}} + 1)$
8.7 $y = (1+x^2)^{\frac{1}{2}}$
8.8 $y = (\cos^2 x + 1)^{\frac{1}{2}}$
8.9 $y = \tan(x^{\frac{1}{3}} + 1)$
8.10 $y = (\sec x + 1)^{-\frac{1}{3}}$

Find the exact values; that is, $\sin^{-1}\frac{\sqrt{2}}{2} = \frac{\pi}{4}$ (not .785).

8.11 $\sin^{-1}\dfrac{1}{2}$
8.12 $\cos^{-1}\dfrac{\sqrt{2}}{2}$
8.13 $\tan^{-1}\dfrac{1}{\sqrt{3}}$
8.14 $\cos^{-1}\dfrac{\sqrt{3}}{2}$
8.15 $\sin^{-1}\left(-\dfrac{\sqrt{3}}{2}\right)$
8.16 $\cos^{-1}\left(-\dfrac{1}{2}\right)$
8.17 $\cos^{-1}(-1)$
8.18 $\sin^{-1}(1)$
8.19 $\sin^{-1}(\sin\pi)$
8.20 $\cos^{-1}\left(\cos\left(-\dfrac{\pi}{3}\right)\right)$

Find $\frac{dy}{dx}$.

8.21 $y = \sin^{-1} x^2$
8.22 $y = \sin^{-1} 3x$
8.23 $y = \sin^{-1}(2x-1)$
8.24 $y = \cos^{-1}\sqrt{x}$
8.25 $y = \cos^{-1}(e^x)$
8.26 $y = \cos^{-1}(x+1)$
8.27 $y = \tan^{-1} 3x$
8.28 $y = \tan^{-1}(x+2)$
8.29 $y = \tan^{-1}(\log x)$

8.30 $y = \tan^{-1}\left(\dfrac{\sin x}{\cos x}\right)$

8.31 (a) Verify that $\frac{d}{dx}\tan^{-1}\frac{1}{x} = -\frac{d}{dx}\tan^{-1}x$

(b) Draw a right triangle which shows that $\tan^{-1}x + \tan^{-1}\frac{1}{x} = \frac{\pi}{2}$.

8.32 Use the chain rule to show that $\frac{d}{dx}(x^{\frac{m}{n}}) = \frac{d}{dx}(x^m)^{\frac{1}{n}} = \frac{m}{n}x^{\frac{m}{n}-1}$.

8.33 Show that the identity $\cos(\sin^{-1}x) = \sqrt{1-x^2}$ follows from the identity $\cos u = \sqrt{1-\sin^2 u}$.

8.34 (a) Verify: $\frac{d}{dx}\frac{1}{a}\tan^{-1}(\frac{x+b}{a}) = \frac{1}{a^2+(x+b)^2}$.

(b) Find y if $\frac{dy}{dx} = \frac{1}{25+(x-2)^2}$.

8.35 Use Figures 8.1 and 8.2 to write the following as algebraic functions of x:

(a) $\tan(\sin^{-1}x)$

(b) $\cos(\tan^{-1}x)$

(c) $\sec(\sin^{-1}x)$

(d) $\tan(\sec^{-1}\sqrt{1+x^2})$

9

Derivatives and Graphs

In this section we see how information about the graph of a function can be obtained from its derivative f', and still more information from the derivative of the derivative, f''.

If $f'(x_0) > 0$, then since this positive slope is the limit of the slopes of segments from $(x_0, f(x_0))$ to nearby points $(x, f(x))$, all such segments must have positive slope. Hence, $f(x) > f(x_0)$ for all x just to the right of x_0, and $f(x) < f(x_0)$ for all x just to the left of x_0.

We say that f has a **local maximum** at x_0, provided $f(x_0) \geq f(x)$ for all x sufficiently close to x_0, and a **local minimum** at x_0, provided $f(x_0) \leq f(x)$ for all x sufficiently close to x_0. From the argument above it is clear that f can have neither a local maximum nor a local minimum at x_0 if $f'(x_0) > 0$ or if $f'(x_0) < 0$. Therefore, if $f(x_0)$ is either a local maximum or a local minimum, then $f'(x_0) = 0$. The condition $f'(x_0) = 0$ is necessary for $f(x_0)$ to be a local maximum or minimum, but not a sufficient condition; f' can be zero at points where f does not have a maximum or minimum. (See Figure 9.1.) If we are concerned only about the values of $f(x)$ for x in some closed interval $[a, b]$, then f can have a local maximum or minimum at a or b without f' being zero. Indeed, if, for example, $f'(a) > 0$, then $f(a)$ is necessarily a local minimum on $[a, b]$.

If $f'(x) > 0$ for all x in some interval, then $f(x)$ is strictly increasing throughout that interval. Similarly, $f(x)$ is strictly decreasing on any interval on which $f'(x) < 0$. These facts are consequences of the following important result:

Mean Value Theorem: *If $f'(x)$ exists for $a \leq x \leq b$, then there is some $c \in (a, b)$ such that*

$$f(b) - f(a) = f'(c)(b - a).$$

The Mean Value Theorem is simply a precise statement of the obvious fact (Figure 9.2) that a graph cannot get smoothly (i.e., no corners) from $(a, f(a))$ to $(b, f(b))$ without pointing in the right direction for at least one point c between a and b. As another interpretation of the Mean Value Theorem, suppose $f(t)$ is the distance your car goes from time $t = a$ to time $t = b$. The average speed over the interval $a \leq t \leq b$ is $(f(b) - f(a))/(b - a)$. The Mean Value Theorem (and common sense) says that there must be a least one point $t = c$ at which the instantaneous speed $f'(c)$ equals the average speed.

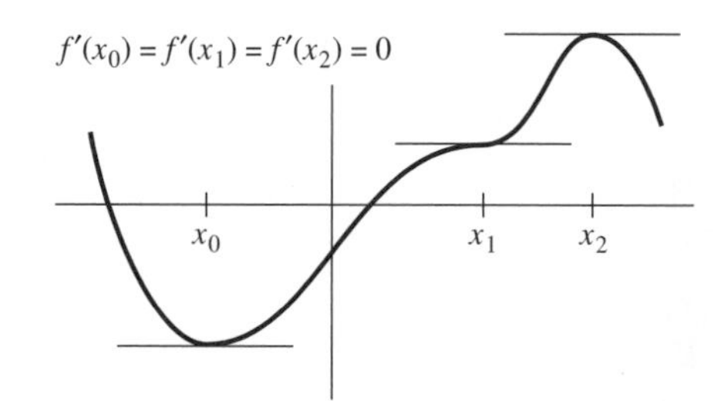

Figure 9.1

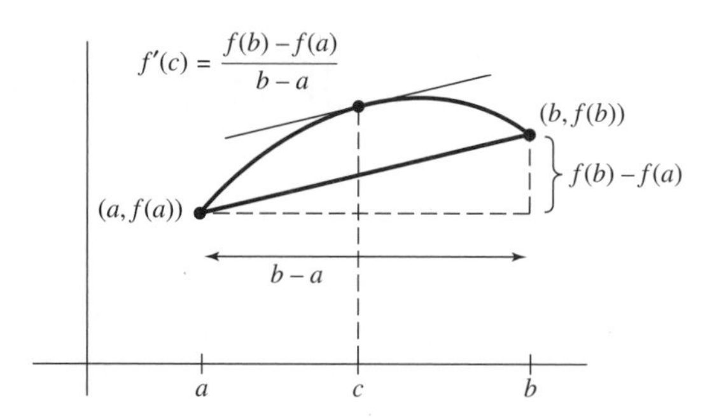

Figure 9.2

One obvious but very important consequence of the Mean Value Theorem is the fact that if $f'(x) \equiv 0$ on some interval, then $f(x)$ is constant on that interval. That is, if x_1 and x_2 are any two points, then $f(x_1) = f(x_2)$ because

$$f(x_2) - f(x_1) = f'(c)(x_2 - x_1) = 0 \cdot (x_1 - x_2) = 0.$$

A corollary of this fact that we will use in integration theory is the following: *If two functions have the same derivative, then they differ by a constant.* That is, if $f'(x) \equiv g'(x)$, and $h(x) = f(x) - g(x)$, then $h'(x) \equiv 0$ so $h(x) = f(x) - g(x)$ is a constant.

If $f'(x) > 0$ for all x in some interval I, then for any interval $[x_1, x_2]$ within I we can apply the Mean Value Theorem to get

$$f(x_2) - f(x_1) = f'(c)(x_2 - x_1).$$

Since $f'(c) > 0$, $f(x_2) > f(x_1)$ and f is increasing in I.

If $f'(x) > 0$ on an interval (x_1, x_0), so $f(x)$ is increasing up to $f(x_0)$, and $f'(x) < 0$ on an interval (x_0, x_2), so $f(x)$ decreases away from x_0, then $f(x_0)$ is a local maximum. Similarly, if $f'(x) < 0$ on (x_1, x_0) and $f'(x) > 0$ on (x_0, x_2), then $f(x_0)$ is a local minimum.

If $f'(x_0) = 0$, the value of the second derivative at x_0 furnishes an easy way to check the sign of $f'(x)$ on either side of x_0. If $f'(x_0) = 0$ and $f''(x_0) > 0$, then $f'(x) > 0$ on some interval to the right of x_0, and $f'(x) < 0$ on some interval to the left of x_0. It follows that $f(x_0)$ is a local minimum if $f'(x_0) = 0$, $f''(x_0) > 0$. Similarly, if $f'(x_0) = 0$ and $f''(x_0) < 0$, then $f(x_0)$ is a local maximum.

The common functions we deal with in calculus—powers, roots, exponentials, logarithms, trigonometric functions, and inverse trigonometric functions—all have derivatives of all orders on any open interval on which the function is defined. The notation for these successive derivatives is as follows: if $y = f(x)$, then

$$\frac{dy}{dx} = f'(x), \ \frac{d^2y}{dx^2} = f''(x), \ \frac{d^3y}{dx^3} = f'''(x), \ \frac{d^4y}{dx^4} = f^{(4)}(x), \ldots.$$

We write $\frac{d}{dx}$ to indicate the derivative of whatever follows. For example,

$$\frac{d}{dx}f(x) = f'(x), \quad \frac{d}{dx}(3x^2 + \sin x) = 6x + \cos x.$$

The notation $\frac{d^n y}{dx^n}$ then indicates the differentiation operator $\frac{d}{dx}$ applied n times, as suggested by the notation $(\frac{d}{dx})^n y = \frac{d^n y}{dx^n}$. The third and higher order derivatives become important later when we discuss power series, but for now we will be concerned only with the first and second derivatives.

EXAMPLE 9.1

Find the local maxima and minima of $f(x) = \frac{1}{6}x^3 - \frac{1}{2}x^2 - \frac{3}{2}x + 2$.

Solution

We know that a local maximum or minimum can only occur where $f'(x) = 0$, so we first find all these so-called **critical points** where $f'(x) = 0$.

$$f(x) = \frac{1}{6}x^3 - \frac{1}{2}x^2 - \frac{3}{2}x + 2,$$

$$f'(x) = \frac{1}{2}x^2 - x - \frac{3}{2};$$

$f'(x) = 0$ if and only if

$$\frac{1}{2}x^2 - x - \frac{3}{2} = 0,$$

$$x^2 - 2x - 3 = 0,$$

$$(x-3)(x+1) = 0,$$

$$x = 3 \text{ or } x = -1.$$

The critical points, where there *might* be a local maximum or local minimum, are $x = -1$ and $x = 3$. We check the second derivative at these points.

$$f''(x) = x - 1,$$

$$f''(-1) = -2 < 0;\ f''(3) = 2 > 0.$$

Since $f'(-1) = 0$, $f''(-1) < 0$, f has a local maximum at $(-1, \frac{17}{6})$. Since $f'(3) = 0$ and $f''(3) > 0$, f has a local minimum at $(3, -\frac{5}{2})$. The graph is shown in Figure 9.3.

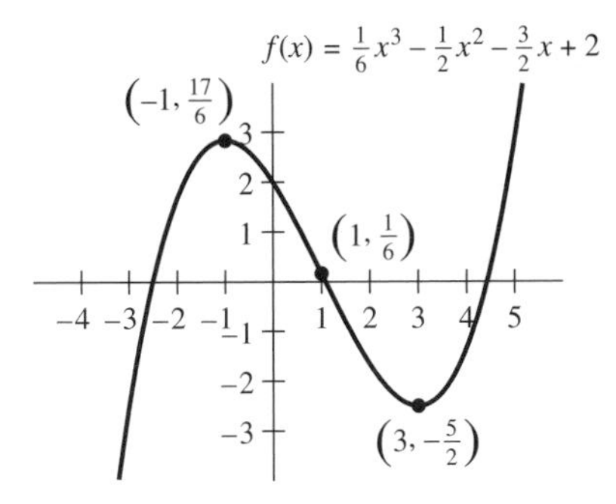

Figure 9.3

If $f''(x) > 0$ on an interval I, then $f'(x)$ is increasing on I and the graph of such a function is called **concave up**. If $f''(x) < 0$ on I, then $f'(x)$ is decreasing on I, and the graph is **concave down**. For the function

$$f(x) = \frac{1}{6}x^3 - \frac{1}{2}x^2 - \frac{3}{2}x + 2$$

of the preceding example, we saw that $f''(x) = x - 1$. Therefore, $f''(x) > 0$ if $x > 1$, so the graph is concave up on $(1, \infty)$, and $f''(x) < 0$ if $x < 1$, so the graph is concave down on $(-\infty, 1)$. A point like $(1, \frac{1}{6})$ where the concavity changes is called a **point of inflexion**. The second derivative is necessarily zero at a point of inflexion, and a cubic that has a local maximum at x_1 and local minimum at x_2 will always have a point of inflexion midway between them at $(x_1 + x_2)/2$ (see Problem 9.16).

The graph of any equation in x and y is some sort of curve in the plane, but this curve need not be the graph of a function. For example, the equation $x^2 + y^2 = 1$ does not define y as a function of x, since there are two values of y for each x in $(-1, 1)$. However, if we restrict our attention to part of the graph—say the top half of $x^2 + y^2 = 1$—we do get the graph of a function (in this case $y = \sqrt{1 - x^2}$). It is easy to find $\frac{dy}{dx}$ for such functions y whose graphs form part of the graph of an equation. We simply differentiate both sides of the equation to get an equation for $\frac{dy}{dx}$. If you differentiate both sides of $x^2 + y^2 = 1$, you get

$$2x + 2y\frac{dy}{dx} = 0,$$

$$\frac{dy}{dx} = -\frac{x}{y}.$$

Notice that the value of $\frac{dy}{dx}$ depends on both y and x as it must, since the slope on the top half of the circle ($y = \sqrt{1 - x^2}$) is the negative of the slope on the bottom half ($y = -\sqrt{1 - x^2}$).

In general, it may not be possible to solve for y in terms of x. Nevertheless, the procedure above, called **implicit differentiation**, will give a value for $\frac{dy}{dx}$ at any point of the curve.

EXAMPLE 9.2

Find $\frac{dy}{dx}$ and $\frac{d^2y}{dx^2}$ at (2, 1) for the function y which is defined implicitly by the equation $2y^3 + 6y = 3x^2 - 4$.

Solution

Notice that it would not be easy to solve this cubic for y in terms of x. Nevertheless, the curve does go through (2, 1), and we can find $\frac{dy}{dx}$ and $\frac{d^2y}{dx^2}$ there:

$$6y^2\frac{dy}{dx} + 6\frac{dy}{dx} = 6x,$$

$$\frac{dy}{dx}(6y^2 + 6) = 6x, \tag{9.1}$$

$$\frac{dy}{dx} = \frac{x}{1 + y^2}.$$

At (2, 1), $\frac{dy}{dx} = 1$. Now differentiate $x/(1 + y^2)$ to find $\frac{d^2y}{dx^2}$.

$$\frac{d^2y}{dx^2} = \frac{(1 + y^2) - x(2y\frac{dy}{dx})}{(1 + y^2)^2}.$$

Substituting 2 for x, 1 for y, and 1 for $\frac{dy}{dx}$ gives the value of $\frac{d^2y}{dx^2}$ at (2, 1):

$$\frac{d^2y}{dx^2} = \frac{(1 + 1) - 2(2 \cdot 1 \cdot 1)}{(1 + 1)^2} = -\frac{1}{2}.$$

PROBLEMS

Find $f'(x)$ and $f''(x)$.

9.1 $f(x) = x^3 - 5x^2 + 3x - 1$

9.2 $f(x) = x^7 - 2x^3 + 3x^{-4}$

9.3 $f(x) = x^{\frac{1}{2}} - x^{-\frac{1}{3}}$

9.4 $f(x) = \dfrac{1}{1 + x^2}$

Find $\frac{dy}{dx}$ and $\frac{d^2y}{dx^2}$.

9.5 $y = e^{x^2}$

9.6 $y = \sin 2x - \cos 3x$

9.7 $y = \sin^{-1} x$

9.8 $y = \tan^{-1} \dfrac{x}{2}$

Find the local maxima and minima and graph the curve.

9.9 $y = x^3 - 3x$

9.10 $y = x^3 + 3x^2 + 4$

9.11 $y = 4x + \dfrac{1}{x}$

9.12 $y = x^3(1 - x)$. *Hint*: If $\frac{d^2y}{dx^2} = 0$ at a critical point, consider $\frac{dy}{dx}$ on either side.

9.13 $y = x^4 + 4x + 3$

9.14 $y = x^3 - 6x^2 + 9x - 2$

9.15 Show that the local maximum of $y = x + \frac{1}{x}$ is less than the local minimum.

9.16 If $f(x)$ is a cubic polynomial that has a local maximum at $x = a$ and a local minimum at $x = b$, then f has an inflexion point at $x = \frac{a+b}{2}$, which is halfway from a to b. *Hint*: If $f(x)$ is cubic, then $f'(x)$ is a quadratic. Moreover, $f'(a) = f'(b) = 0$, so $f'(x) = k(x - a)(x - b)$.

9.17 Find the dimensions of the rectangle of largest area that has a perimeter of 40 in.

9.18 A pan is to be made from a 12″ square piece of tin by cutting squares out of the corners and folding up the sides. What is the maximum volume of such a pan?

9.19 What is the ratio of height (h) to radius (r) if a can of volume V is to have minimum total area? ($A = 2\pi r^2 + 2\pi rh$; $V = \pi r^2 h$. V is given, so $h = V/\pi r^2$.)

9.20 (i) What is the largest product of two positive integers whose sum is 10; that is, what is the largest of the products $1 \cdot 9, 2 \cdot 8, 3 \cdot 7, 4 \cdot 6, 5 \cdot 5$?

(ii) What is the largest product of two positive numbers whose sum is a given number B? (N.B. *numbers*, not integers.)

9.21 Find the radius x and height h of the largest cylinder which can be inscribed in a cone of radius R and height H.

9.22 Find the point (x, y) on the line $y = 3 - \frac{3}{2}x$ which is closest to $(0, 1)$. *Hint*: Let s be the square of the distance and find (x, y) on the line so s is minimum.

9.23 Find the point (x, y) on the curve $y^2 = 2(x - 1)$ which is closest to the origin.

9.24 Find the shortest ladder that will fit over a fence h feet high to a wall b feet behind the fence. *Hint*: Let θ be the angle the ladder makes with the ground. Show that the length is $\ell = \frac{h}{\sin\theta} + \frac{b}{\cos\theta}$ and $\frac{d\ell}{d\theta} = 0$ when $\tan\theta = \frac{h^{\frac{1}{3}}}{b^{\frac{1}{3}}}$.

9.25 The cost per hour of running a cargo boat is proportional to the cube of the speed through the water. Find the speed through the water, v, which minimizes the cost of a given trip upriver against a a 5 mph current. *Hint*: The speed over the ground is $v - 5$, so the time of the trip is proportional to $\frac{1}{(v-5)}$.

9.26 A piece of string 100 in long is cut into two pieces. One piece, of length x, is formed into a circle, and the rest into a square. Let $A(x)$ be the sum of the two areas. Find x so that $A(x)$ is maximum and find x so that $A(x)$ is minimum. *Hint*: There is one x_0 between 0 and 100 such that $A'(x_0) = 0$, and you can easily tell whether $A(x_0)$ is a maximum or minimum. To find the other extreme value, consider how $A(x)$ behaves on $[0, x_0]$ and on $[x_0, 100]$. Remember that $x = 0$ (all circle) and $x = 100$ (all square) are possibilities.

10

Following the Tangent Line

The tangent line to $y = f(x)$ at x_0 has the equation

$$y = f(x_0) + f'(x_0)(x - x_0). \tag{10.1}$$

Consequently, if you know $f(x_0)$ and $f'(x_0)$, you can calculate exactly the value of y on the tangent line for any given x. If x is close to x_0, then the graph of f will be close to the tangent line, and we can use the value of y from (10.1) to approximate the value of $f(x)$. In this section, we examine several ways this approximation can be used.

Suppose we want a rough approximation for $\sqrt{4.04}$ or $\sqrt{3.99}$ and the calculator is not at hand. If $f(x) = \sqrt{x}$, then $f'(x) = 1/2\sqrt{x}$ and $f'(4) = \frac{1}{4}$. We know that $\sqrt{4} = 2$ and the slope of the tangent at $x = 4$ is $\frac{1}{4}$. Therefore, $\sqrt{x}$ changes from $\sqrt{4}$ by about $\frac{1}{4}$ the change in x from 4. Thus,

$$\sqrt{4.04} \doteq 2 + \frac{1}{4}(.04) = 2.01 \text{ and}$$

$$\sqrt{3.99} \doteq 2 + \frac{1}{4}(-.01) = 2 - .003 = 1.997.$$

The notation $f'(x)\,dx$ is frequently used for the change on the tangent line corresponding to a change dx from x, and dy is then defined by $dy = f'(x)\,dx$ to correspond to the $\frac{dy}{dx} = f'(x)$ notation. In the preceding example we would have $f(x) = \sqrt{x}$, $x = 4$, $dx = .04$, giving

$$dy = \frac{1}{2\sqrt{4}}(.04) = .01.$$

The expression $f'(x)\,dx$ is called the **differential of f at x**.

As another example of the use of the differential, suppose that your business is stamping out circular pieces of sheet metal. The buyer cares not only about the accuracy in the radius, but also about the accuracy in the weight of each piece. You therefore need to know how the relative error in weight, $\frac{dW}{W}$, depends on the relative error, $\frac{dr}{r}$, in the radius. If ρ is the density (ounces per square inch, say), then $W = \rho\pi r^2$ and $dW = 2\rho\pi r dr$; hence,

$$\frac{dW}{W} = \frac{2\rho\pi r dr}{\rho\pi r^2} = \frac{2dr}{r}.$$

The relative error in weight is twice the relative error in the radius.

The number of years it takes to double your money at a compound interest rate r is $n = \log 2/\log(1+r)$. If $f(x) = \log(1+x)$, then $f'(x) = 1/(1+x)$. Hence, $f(0) = \log 1 = 0$ and $f'(0) = 1$. The change in $\log(1+x)$ from $x = 0$ is approximately $1 \cdot x$. Thus, $\log(1+r) \doteq r$. For example, at 5% ($r = .05$) and using .70 for a crude approximation to $\log 2$, we get

$$n = \frac{.70}{.05} = \frac{70}{5} \doteq 14 \text{ years.}$$

This is the businessman's rule of thumb: The number of years to double your money at interest $p\%$ is $70/p$. The formula for tripling your money is $n = \log 3/\log(1+r) \doteq 1.10/r$. For example, at $10\% (r = .10)$, $n = 1.10/.10 = 11$ years.

Tangential approximation can also be used to evaluate certain nonobvious limits of the form $0/0$ using the following result:

l'Hospital's Rule: If $\lim_{x\to a} f(x) = f(a) = 0$ *and* $\lim_{x\to a} g(x) = g(a) = 0$, *and* $g'(a) \neq 0$, *then* $\lim_{x\to a} \frac{f(x)}{g(x)} = \lim_{x\to a} \frac{f'(x)}{g'(x)} = \frac{f'(a)}{g'(a)}$.

To verify this, notice that if we define $\varepsilon_1(x)$ by

$$\frac{f(x) - f(a)}{x - a} - f'(a) = \varepsilon_1(x),$$

then $\varepsilon_1(x) \longrightarrow 0$ as $x \longrightarrow a$ since $f'(a)$ is the limit of the difference quotient $(f(x) - f(a))/(x - a)$. Therefore, since $f(a) = 0$,

$$f(x) = f'(a)(x - a) + \varepsilon_1(x)(x - a)$$

where $\varepsilon_1(x) \longrightarrow 0$ as $x \longrightarrow a$. Similarly,

$$g(x) = g'(a)(x - a) + \varepsilon_2(x)(x - a)$$

where $\varepsilon_2(x) \longrightarrow 0$ as $x \longrightarrow a$. Therefore, for $x \neq a$,

$$\frac{f(x)}{g(x)} = \frac{f'(a)(x - a) + \varepsilon_1(x)(x - a)}{g'(a)(x - a) + \varepsilon_2(x)(x - a)}$$

$$= \frac{f'(a) + \varepsilon_1(x)}{g'(a) + \varepsilon_2(x)} \longrightarrow \frac{f'(a)}{g'(a)}.$$

To apply l'Hospital's Rule, first check that $f(a) = g(a) = 0$, and then write

$$\lim_{x\to a} \frac{f(x)}{g(x)} = \lim_{x\to a} \frac{f'(x)}{g'(x)}.$$

For example,

$$\lim_{x\to 0} \frac{e^x - 1}{x} = \lim_{x\to 0} \frac{e^x}{1} = 1;$$

$$\lim_{x\to 0} \frac{1 - \cos x}{x} = \lim_{x\to 0} \frac{\sin x}{1} = 0.$$

If $\lim_{x\to a} f'(x)/g'(x)$ also has the form $\frac{0}{0}$, then apply l'Hospital's Rule again:

$$\lim_{x\to 0} \frac{1 - \cos x}{x^2} = \lim_{x\to 0} \frac{\sin x}{2x}$$

$$= \lim_{x\to 0} \frac{\cos x}{2} = \frac{1}{2}.$$

We can also use tangential approximation to find numerical estimates for the roots of equations. In geometric terms, to solve the equation $f(x) = 0$ means to find the numbers r where the curve $y = f(x)$ crosses the x-axis. If x_1 is a first approximation to a root r, then we get a better approximation, x_2, by following the tangent line back to the x-axis as shown in Figure 10.1.

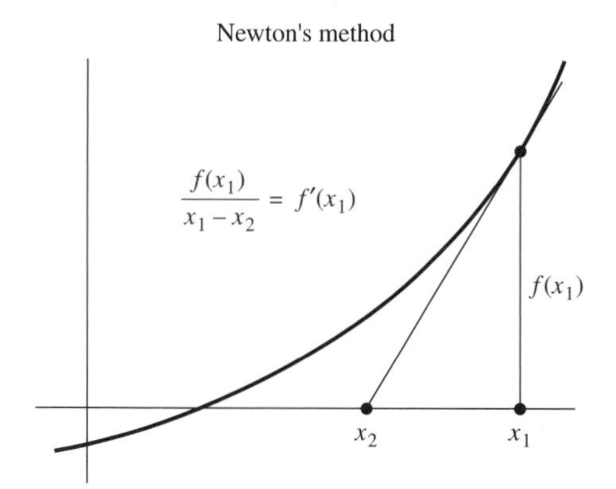

Figure 10.1

The line from $(x_2, 0)$ to $(x_1, f(x_1))$ has slope $f'(x_1)$, so

$$\frac{f(x_1)}{x_1 - x_2} = f'(x_1),$$

$$x_1 - x_2 = \frac{f(x_1)}{f'(x_1)}, \tag{10.2}$$

$$x_2 = x_1 - \frac{f(x_1)}{f'(x_1)}.$$

The number x_2 determined by (10.2) will generally be a better approximation to the root r than x_1. We can then use the same formula to get a better approximation, x_3, from x_2:

$$x_3 = x_2 - \frac{f(x_2)}{f'(x_2)}.$$

This technique is called **Newton's method**.

EXAMPLE 10.1

Use Newton's method to approximate $\sqrt{3}$.

Solution

We let $f(x) = x^2 - 3$ and search for a root of the equation $f(x) = 0$. Since $f(1) < 0$ and $f(2) > 0$, there will be a root between 1 and 2, and we start with $x_1 = 1.5$. Since $f'(x) = 2x$, we have

$$\begin{aligned} x_2 &= 1.5 - \frac{(1.5)^2 - 3}{2(1.5)} \\ &= 1.5 - \frac{(2.25 - 3)}{3.0} \\ &= 1.5 + \frac{.75}{3} = 1.75. \end{aligned}$$

Now calculate x_3:

$$\begin{aligned} x_3 &= 1.75 - \frac{(1.75)^2 - 3}{2(1.75)} \\ &= 1.75 - \frac{.0625}{3.5} \\ &= 1.73214. \end{aligned}$$

The five decimal place approximation to $\sqrt{3}$ is

$$\sqrt{3} \doteq 1.73205,$$

so x_3 is off by only .00009, close enough for government work. Notice that since x_1 and x_2 are only rough approximations, it is pointless to do the arithmetic with many significant figures. Use more significant figures for x_3, x_4, and so on, and you will be able to see the accuracy improving as the first few decimals remain the same in subsequent approximations.

PROBLEMS

Use differentials to approximate the following.

10.1 $\sin 29^\circ \left(\text{i.e., } \sin\left(\frac{\pi}{6} - \frac{\pi}{180}\right)\right)$

10.2 $\sqrt{4.02}$

10.3 $e^{0.2}$

10.4 $\tan^{-1} 1.04$

10.5 $\log 1.002$

10.6 2.01^3

10.7 $30^{\frac{1}{5}}$

10.8 Estimate $f(3.1)$, given that $f(3) = 25$ and $f'(x) = \sqrt{1+x}$.

10.9 Find the relative error in the weight of a ball bearing ($W = \frac{4}{3}\rho\pi r^3$) in terms of the relative error in the radius measurement.

Find the following limits.

10.10 $\lim\limits_{x\to 0} \frac{e^x - 1}{3x}$

10.11 $\lim\limits_{x\to 0} \frac{e^x - 1 - x}{x^2}$

10.12 $\lim\limits_{x\to 1} \frac{\log x}{2x - 2}$

10.13 $\lim\limits_{x\to 0} \frac{\log(1+x)}{x}$

10.14 $\lim\limits_{x\to \frac{\pi}{2}} \frac{\cos x}{x^2 - \frac{\pi^2}{4}}$

10.15 $\lim\limits_{x\to 0} \frac{1 - \cos x}{x \sin x}$

10.16 $\lim\limits_{x\to 2} \frac{2x - 4}{\sqrt{x^2 - 4}}$

10.17 $\lim\limits_{x\to 0} \frac{\log(1+x)}{\sqrt{x}}$

Use Newton's method as indicated.

10.18 Approximate $\sqrt{5}$. Use $f(x) = x^2 - 5$, with $x_1 = 2.5$. Find x_3.

10.19 Approximate $\sqrt[3]{3}$. Use $x_1 = 1.5$ and find x_3.

10.20 Approximate the root of $x^3 + x - 1 = 0$ which lies between 0 and 1. Use $x_1 = .5$ and continue until you are sure of three decimal places.

10.21 (i) Show that $xe^x - 1 = 0$ has exactly one root. *Hint*: If $f(x) = xe^x - 1$, then $f(x) < 0$ if $x < 0$, and $f'(x) > 0$ if $x > 0$.

(ii) Approximate the root of $xe^x - 1 = 0$ to three decimal places.

11

The Indefinite Integral

In many applications, the information about the dependent variable is given in terms of its derivative or derivatives. For example, if an object falls to the earth under the influence of gravity, Newton's law says that the acceleration is a constant. If s is the distance the object falls in time t, and v and a are its velocity and acceleration of time t, then

$$v = \frac{ds}{dt}, \quad a = \frac{dv}{dt} = \frac{d^2s}{dt^2}. \tag{11.1}$$

Hence, Newton's law in mathematical terms is

$$\frac{d^2s}{dt^2} = g, \tag{11.2}$$

where g is a constant that depends on the units used. (If s is measured in feet and t in seconds, then g is approximately 32 ft/sec^2.)

If we are given an expression like (11.2) which describes the derivatives of s, then naturally we want to be able to convert it into an explicit formula for s. This process is called (indefinite) integration. We start with

$$\frac{d^2s}{dt^2} = \frac{dv}{dt} = g.$$

Of course, $\frac{d}{dt}(gt) = g$, and we know that any other function, v, which has this same derivative, g, must differ from gt by a constant. Therefore,

$$v = gt + c_1$$

for some constant c_1. Now use $\frac{ds}{dt} = v$, so

$$\frac{ds}{dt} = gt + c_1.$$

Clearly, $\frac{1}{2}gt^2 + c_1 t$ has derivative $gt + c_1$, so

$$s = \frac{1}{2}gt^2 + c_1 t + c_2$$

for some constant c_2. The constant c_2 is the value of s when $t = 0$ (the initial value of s), and the constant c_1 is the value of $v = \frac{ds}{dt}$ when $t = 0$ (the initial velocity).

Now we will systematize this process of going from the derivative to the function. If $F'(x) = f(x)$, then $F(x)$ is called an **antiderivative** or **indefinite integral** of $f(x)$, and we indicate this with the notation

$$F(x) = \int f(x)\,dx + c,$$

where c is understood to be an arbitrary constant. The rules for integration are equivalent to the rules for differentiation, since

$$\int f(x)\,dx = F(x) \text{ means } \frac{d}{dx}F(x) = f(x).$$

Here are some of the differentiation formulas we have, with the corresponding integration formulas. In writing integration formulas, we will suppress the arbitrary additive constant and understand that a formula like $\int x\,dx = \frac{1}{2}x^2$ means that $\frac{1}{2}x^2$ is one of the infinitely many integrals of x, the rest all having the form $\frac{1}{2}x^2 + c$.

$$\frac{d}{dx}x^n = nx^{n-1}; \quad \int x^n\,dx = \frac{1}{n+1}x^{n+1} \qquad (n \neq -1);$$

$$\frac{d}{dx}\sin x = \cos x; \quad \int \cos x\,dx = \sin x;$$

$$\frac{d}{dx}\cos x = -\sin x; \quad \int \sin x\,dx = -\cos x;$$

$$\frac{d}{dx}e^x = e^x; \quad \int e^x\,dx = e^x;$$

$$\frac{d}{dx}\log x = \frac{1}{x}; \quad \int \frac{1}{x}\,dx = \log x;$$

$$\frac{d}{dx}\sin^{-1} x = \frac{1}{\sqrt{1-x^2}}; \quad \int \frac{dx}{\sqrt{1-x^2}} = \sin^{-1} x;$$

$$\frac{d}{dx}\tan^{-1} x = \frac{1}{1+x^2}; \quad \int \frac{dx}{1+x^2} = \tan^{-1} x.$$

Notice that integration is a linear operation because differentiation is linear. That is,

$$\int (f(x) + g(x))\,dx = \int f(x)\,dx + \int g(x)\,dx,$$

$$\int af(x)\,dx = a\int f(x)\,dx.$$

For example,

$$\int (3\cos x + 5x)\,dx = 3\int \cos x\,dx + 5\int x\,dx$$
$$= 3\sin x + \frac{5}{2}x^2 + c.$$

EXAMPLE 11.1

Suppose that at every point (x, y) of a curve the slope of the curve is $e^x - x$, and the curve passes through $(0, 2)$. Find y in terms of x.

Solution

We are given the differential equation

$$\frac{dy}{dx} = e^x - x,$$

with initial condition $y(0) = 2$. Hence

$$\begin{aligned} y &= \int (e^x - x)\,dx \\ &= e^x - \frac{1}{2}x^2 + c. \end{aligned}$$

Since $y(0) = 2$, we must have

$$\begin{aligned} 2 &= e^0 - \frac{1}{2}\cdot 0 + c, \\ c &= 1. \end{aligned}$$

Finally,

$$y = e^x - \frac{1}{2}x^2 + 1.$$

The integration process frequently involves some juggling of constants. Consider

$$\int \sin 2x\,dx.$$

We notice that

$$\frac{d}{dx}(-\cos 2x) = 2\sin 2x,$$

so $-\cos 2x$ is almost the answer, with an extra factor of 2 resulting from the chain rule. We write

$$\int \sin 2x\,dx = \frac{1}{2}\int (\sin 2x)(2)\,dx,$$

and notice that the last integral has the form

$$\frac{1}{2}\int \sin u\,du = -\frac{1}{2}\cos u,$$

where $u = 2x$. Thus, with $u = 2x$

$$\begin{aligned} \int \sin 2x\,dx &= \frac{1}{2}\int (\sin 2x)2\,dx \\ &= \frac{1}{2}\int \sin u\,du \\ &= -\frac{1}{2}\cos u \\ &= -\frac{1}{2}\cos 2x. \end{aligned}$$

This technique is called making a u-**substitution**. Here are some more examples:

(i) $\int e^{3x}dx.$

Make the substitution $u = 3x$, $du = 3\,dx$, $dx = \frac{1}{3}\,du$, so

$$\int e^{3x}\,dx = \int e^u \frac{1}{3}\,du$$
$$= \frac{1}{3}\int e^u\,du$$
$$= \frac{1}{3}e^u$$
$$= \frac{1}{3}e^{3x}.$$

(ii) $\int x\cos(4x^2)\,dx$.
Let $u = 4x^2$, $du = 8x\,dx$, $x\,dx = \frac{1}{8}\,du$. Then

$$\int x\cos x^2\,dx = \int \cos u\left(\frac{1}{8}\right)\,du$$
$$= \frac{1}{8}\int \cos u\,du$$
$$= \frac{1}{8}\sin u$$
$$= \frac{1}{8}\sin 4x^2.$$

(iii) $\int \sqrt{1-x^2}\,x\,dx$.

In all these examples, the key is to recognize the chain rule in reverse. Here that means noticing that except for a constant, x is the derivative of $1-x^2$. Hence, we let $u = 1-x^2$, $du = -2x\,dx$, $dx = -\frac{1}{2}\,du$, so

$$\int \sqrt{1-x^2}x\,dx = \int \sqrt{u}\left(-\frac{1}{2}\right)\,du$$
$$= -\frac{1}{2}\int u^{\frac{1}{2}}\,du$$
$$= \left(-\frac{1}{2}\right)\frac{2}{3}u^{\frac{3}{2}}$$
$$= -\frac{1}{3}(1-x^2)^{\frac{3}{2}}.$$

(iv) $\int \frac{3x}{4+x^2}\,dx$.
If $u = 4+x^2$, then $du = 2x\,dx$, and $3x\,dx = \frac{3}{2}\,du$. Therefore,

$$\int \frac{3x}{4+x^2}\,dx = \int \frac{\frac{3}{2}du}{u}$$
$$= \frac{3}{2}\int \frac{du}{u}$$
$$= \frac{3}{2}\log u$$
$$= \frac{3}{2}\log(4+x^2).$$

In the last example we used the formula $\int \frac{du}{u} = \log u$, which is not quite complete.

Recall that $\log x$ is defined only for $x > 0$, and consequently $\frac{d}{dx}\log x = \frac{1}{x}$ only makes sense for $x > 0$. However, $\log|x|$ makes sense for all $x \neq 0$, and it is easy to see (Problem 11.34) that

$$\frac{d}{dx}\log|x| = \frac{1}{x} \tag{11.3}$$

for all $x \neq 0$. Consequently, the proper integration version of (11.3) is

$$\int \frac{dx}{x} = \log|x|.$$

For example, the curve y such that $\frac{dy}{dx} = \frac{1}{x}$ and $y(-e) = 3$ is determined by:

$$y = \int \frac{1}{x}\,dx = \log|x| + c.$$

Since $y(-e) = 3$,

$$\begin{aligned} 3 &= \log|-e| + c \\ &= \log e + c \\ &= 1 + c, \end{aligned}$$

so $c = 2$, and $y = \log|x| + 2$.

PROBLEMS

Find the indefinite integrals.

11.1 $\int x^3\,dx$

11.2 $\int x^{-4}\,dx$

11.3 $\int x^{\frac{7}{3}}\,dx$

11.4 $\int \frac{1}{\sqrt{x}}\,dx$

11.5 $\int 5x^{-\frac{3}{2}}\,dx$

11.6 $\int 7x^{\frac{3}{4}}\,dx$

11.7 $\int \left(2\sqrt{x} + 3x^3\right)\,dx$

11.8 $\int \frac{x^2 + 5x^4}{x}\,dx$

11.9 $\int \sin 3x\,dx$

11.10 $\int \cos 5x\,dx$

11.11 $\int \left(\sin \frac{1}{2}x - 3\cos 2x\right)\,dx$

11.12 $\int x^2 \cos x^3\,dx$

11.13 $\int \sin(1 + x)\,dx$

11.14 $\int x\cos(1 + x^2)\,dx$

11.15 $\int \sqrt{4 + x^3}x^2\,dx$

11.16 $\int (1+x^2)^{10} x\,dx$

11.17 $\int (3x + 5e^{4x})\,dx$

11.18 $\int (1+e^x)^2\,dx$ *Hint*: Square the binomial.

11.19 $\int xe^{x^2}\,dx$

11.20 $\int e^x \cos e^x\,dx$

11.21 $\int \frac{x+3}{x}\,dx$

11.22 $\int \frac{2x^2}{3+x^3}\,dx$

11.23 $\int \frac{2x+1}{x^2+x+5}\,dx$

11.24 $\int \frac{dx}{\sqrt{1-4x^2}}$

11.25 $\int \frac{dx}{\sqrt{4-x^2}} = \int \frac{dx}{2\sqrt{1-\left(\frac{x}{2}\right)^2}}$

11.26 $\int \frac{5dx}{1+9x^2}$

11.27 $\int \frac{dx}{9+x^2} = \int \frac{dx}{9\left(1+\left(\frac{x}{3}\right)^2\right)}$

11.28 $\int \frac{dx}{x+5}$

Find the function y that satisfies the given conditions.

11.29 $\frac{dy}{dx} = 3x^2 + x - 7,\ y(1) = 2$

11.30 $\frac{dy}{dx} = e^x + \frac{1}{1+x},\ y(0) = 2$

11.31 $\frac{dy}{dx} = \frac{1}{1+x^2},\ y(1) = 0$

11.32 $\frac{dy}{dx} = \frac{1}{\sqrt{1-x^2}},\ y(0) = 5$

11.33 Suppose a car's locked brakes provide a constant negative acceleration ($\frac{dv}{dt} = -k$) which will stop the car going 60 mph (88 ft/sec) in 4 seconds. What is k?

11.34 Show that $\frac{d}{dx} \log|x| = \frac{1}{x}$. *Hint*: If $x > 0$, there is nothing to show, so assume $x < 0$ and $\log|x| = \log(-x)$.

12

The Definite Integral

The definite integral we consider in this section has many interpretations and many applications. We start with the simple geometric idea of the area under a curve. Let $f(x)$ be a positive function defined on some interval $[a, b]$. We want to find the area A between the x-axis and the curve $y = f(x)$ for $a \leq x \leq b$. To approximate the area A, slice up the region under the curve into narrow near-rectangles whose sides are the lines $x = x_i$, for points $x_0, x_1, x_2, \ldots, x_n$ between a and b (Figure 12.1). The area of the slice between $x = x_{i-1}$ and $x = x_i$ is approximately $f(c_i)(x_i - x_{i-1})$ for any point c_i in $[x_{i-1}, x_i]$, since the values of $f(x)$ will not vary much in a small interval $[x_{i-1}, x_i]$. The sum of the areas of the small slices is the following **Riemann sum** for f:

$$\sum_{i=1}^{n} f(c_i)(x_i - x_{i-1}). \tag{12.1}$$

For a continuous function, these sums will approach a limit as $\max(x_i - x_{i-1}) \longrightarrow 0$, and this limit is the **definite integral** of f over $[a, b]$, denoted $\int_a^b f(x)\,dx$. This integral is what we define to be the area under the positive function f.

Now let $F(x)$ be any indefinite integral of $f(x)$; that is, $F'(x) = f(x)$. Apply the Mean Value Theorem to $F(x)$ over each subinterval $[x_{i-1}, x_i]$, and for each i choose $c_i \in [x_{i-1}, x_i]$ such that

$$\begin{aligned} F(x_i) - F(x_{i-1}) &= F'(c_i)(x_i - x_{i-1}) \\ &= f(c_i)(x_i - x_{i-1}). \end{aligned} \tag{12.2}$$

Notice that if we add all the terms on the left of (12.2), the intermediate terms all cancel, so

$$\begin{aligned} \sum_{i=1}^{n}(F(x_i) - F(x_{i-1})) &= F(x_n) - F(x_0) \\ &= F(b) - F(a). \end{aligned} \tag{12.3}$$

Hence, from (12.2) and (12.3) we have

$$F(b) - F(a) = \sum_{i=1}^{n} f(c_i)(x_i - x_{i-1}), \tag{12.4}$$

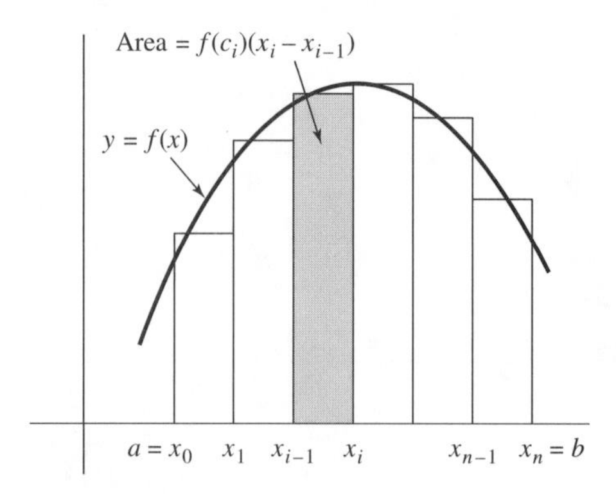

Figure 12.1

where $F(x)$ is any indefinite integral of $f(x)$, and the c_i are appropriately chosen. Since the Riemann sums on the right of (12.4) converge to the definite integral as $\max_i(x_i - x_{i-1}) \longrightarrow 0$,

$$F(b) - F(a) = \int_a^b f(x)\,dx$$

for any F such that $F'(x) = f(x)$. We use the convenient notation

$$F(x)\Big]_a^b = F(b) - F(a), \tag{12.5}$$

so

$$\int_a^b f(x)\,dx = F(x)\Big]_a^b = F(b) - F(a) \tag{12.6}$$

for any antiderivative $F(x)$. The variable x on the left of (12.6) is a dummy variable that could be replaced by any other variable without changing the meaning. Thus

$$\int_a^b f(x)\,dx = \int_a^b f(t)\,dt = \int_a^b f(y)\,dy = \cdots. \tag{12.7}$$

EXAMPLE 12.1

Find the area bounded by the coordinate axes, the curve $y = \frac{1}{2}e^x$, and the line $x = 2$.

Solution

The first step in any area problem (and this is important) is to graph the curves (Figure 12.2). The area is

$$\int_0^2 \frac{1}{2}e^x\,dx = \frac{1}{2}e^x\Big]_0^2 = \frac{1}{2}[e^2 - 1] \doteq 3.19.$$

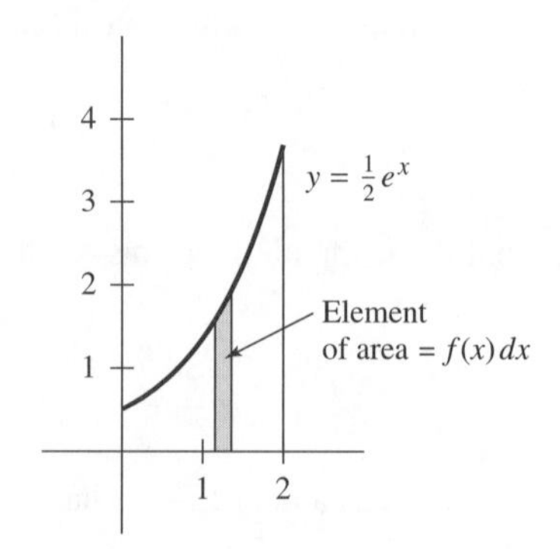

Figure 12.2

The notation $\int_a^b f(x)\,dx$ is meant to suggest the limit of sums of terms of the form $f(x)\,dx$. For area problems we regard $f(x)$ and dx as the height and width of a typical small rectangular slice of the region, and this provides a good mnemonic device for setting up area integrals. For example, let us find the area between the parabolas $y = x^2 - 2$ and $y = 4x - x^2 - 2$. The first step (and this is still important) is to draw the figure. To graph the second parabola we complete the square:

$$\begin{aligned} 4x - x^2 - 2 &= -(x^2 - 4x + 4) - 2 + 4 \\ &= 2 - (x - 2)^2. \end{aligned}$$

The parabola has the line $x = 2$ as its axis, and its vertex is (2,2) (Figure 12.3).

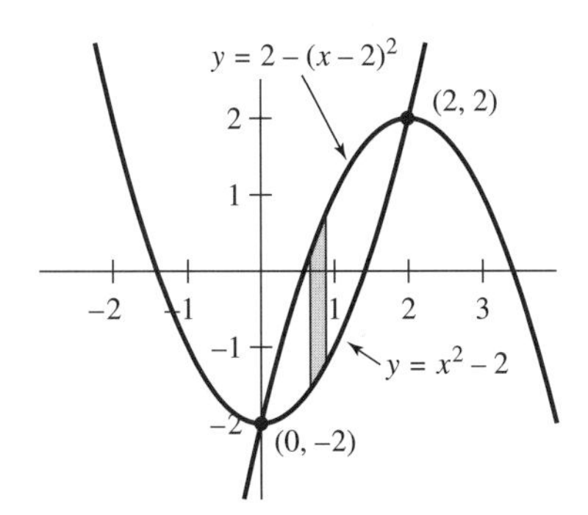

Figure 12.3

A typical slice of the region (and we get this information from the figure) is a "rectangle" whose height is the difference of the y-values on the two curves; namely,

$$(4x - x^2 - 2) - (x^2 - 2) = 4x - 2x^2. \tag{12.8}$$

The x-range is from $x = 0$ to $x = 2$, which we see from the graphs. If the graphs were not quite so nice, then we would have to solve the two equations simultaneously to find the intersections of the curves. The area is, from (12.8),

$$\begin{aligned} A &= \int_0^2 (4x - 2x^2)\, dx \\ &= 2x^2 - \frac{2}{3}x^3 \Big]_0^2 \\ &= 8 - \frac{16}{3} = \frac{8}{3}. \end{aligned}$$

The same slicing technique will work for curves given in the form $x = f(y)$. For example, the area bounded by the y-axis and the parabola

$$x = 2 + y - y^2 = \frac{9}{4} - \left(y - \frac{1}{2}\right)^2$$

(Figure 12.4) is given by

$$\begin{aligned} A &= \int_{-1}^2 (2 + y - y^2)\, dy \\ &= 2y + \frac{1}{2}y^2 - \frac{1}{3}y^3 \Big]_{-1}^2 \\ &= 4 + 2 - \frac{8}{3} - \left(-2 + \frac{1}{2} + \frac{1}{3}\right) \\ &= 8 - \frac{9}{3} - \frac{1}{2} = \frac{9}{2}. \end{aligned}$$

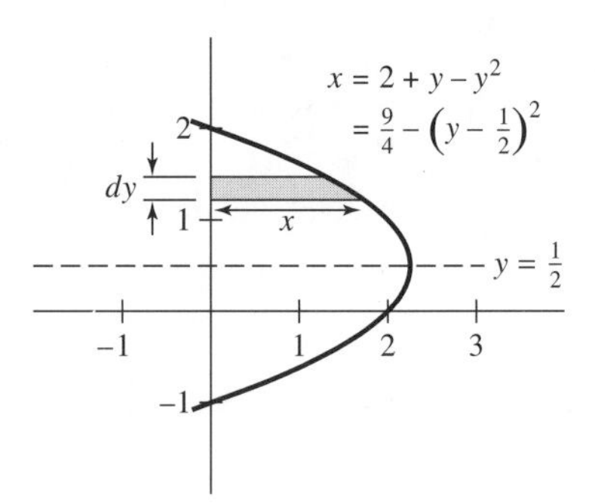

Figure 12.4

In our definition of $\int_a^b f(x)\,dx$, we assumed that $a < b$. It is clear from the definition that if $a < b < c$, then

$$\int_a^b f(x)\,dx + \int_b^c f(x)\,dx = \int_a^c f(x)\,dx. \tag{12.9}$$

It is convenient to make the following additional agreements: if $a < b$, we define

$$\int_b^a f(x)\,dx = -\int_a^b f(x)\,dx,$$

and for any a we define

$$\int_a^a f(x)\,dx = 0.$$

With these conventions, the identity (12.9) holds for any three numbers a, b, c; for example,

$$\int_5^3 f(x)\,dx + \int_3^6 f(x)\,dx = \int_5^6 f(x)\,dx.$$

PROBLEMS

Evaluate the definite integrals.

12.1 $\int_0^2 (x^2 - 2x + 7)\,dx$

12.2 $\int_{-1}^1 (3x^2 + x^5)\,dx$

12.3 $\int_0^4 (2\sqrt{x} + 5x)\,dx$

12.4 $\int_{-3}^{-2} \left(x^{\frac{2}{3}} + \frac{1}{x}\right) dx$

12.5 $\int_1^7 \sqrt{2 + 2y}\,dy$

12.6 $\int_1^{\pi} \cos\left(\frac{1}{2}y\right) dy$

12.7 $\int_0^1 x\sqrt{1 - x^2}\,dx$

12.8 $\int_0^{\frac{\pi}{2}} \sin 2x\,dx$

12.9 $\int_0^1 xe^{x^2}\,dx$

12.10 $\displaystyle\int_0^1 \frac{dx}{1+x^2}$

12.11 $\displaystyle\int_0^{\frac{1}{2}} \frac{dx}{\sqrt{1-x^2}}$

12.12 $\displaystyle\int_0^1 \frac{x\,dx}{1+x^2}$

Graph the region and find the area bounded by the following curves.

12.13 The x-axis and the curve $y = 9 - x^2$.

12.14 The x-axis and the one arch of the curve $y = \sin x$ (e.g., $0 \le x \le \pi$).

12.15 The curves $y = 4 - x^2$ and $y = x + 2$.

12.16 The curves $y = 1 - x^2$ and $y = x^2 - 1$.

12.17 The y-axis and the curve $x = 2y - y^2$.

12.18 The y-axis, the curve $x = y^3$, and the line $y = 1$.

12.19 The curves $y = \sin x$ and $y = \frac{2}{\pi}x$ for $0 \le x \le \frac{\pi}{2}$.

12.20 The coordinate axes, the curve $y = 2x^2 + 4x + 3$, and the line $x = 1$.

12.21 Show that $\frac{d}{dx}\int_a^x f(u)\,du = f(x)$. *Hint*: Let $F'(x) = f(x)$ so $\int_a^x f(u)\,du = F(x) - F(a)$.

12.22 Use Problem 12.21 to calculate the following.

(i) $\displaystyle\frac{d}{dx}\int_0^x e^{t^2}\,dt$

(ii) $\displaystyle\frac{d}{dx}\int_1^x \sqrt{1+u^3}\,du$

(iii) $\displaystyle\frac{d}{dx}\int_\pi^x \sin\theta^2 d\theta$

(iv) $\displaystyle\frac{d}{dx}\int_x^0 \log(1+s^2)\,ds$

12.23 Use Problem 12.21 and the chain rule to evaluate $\frac{d}{dx}\int_a^{g(x)} f(u)\,du$. *Hint*: Let $F(x) = \int_a^x f(u)\,du$, so the problem is to evaluate $\frac{d}{dx}F(g(x))$.

13

Work, Volume, Force

In this chapter we consider some other applications of the definite integral. First, we examine the work done by a force acting over a given distance—for example, the work done pushing a car up a slope or lifting a bucket of water. For a constant force F acting in the direction of the motion, work is simply force times distance:

$$W = F \cdot d.$$

If the force is not constant (the car is pushed up a slope of increasing steepness, or the bucket is leaking water as it's lifted), then we are led to a definite integral.

Suppose the force applied at a point x is given by the nonconstant function $f(x)$. We divide up the interval $[a, b]$ over which the force acts into small subintervals $[x_{i-1}, x_i]$ on which the force is nearly constant. If $c_i \in [x_{i-1}, x_i]$, then $f(x)$ will differ little from $f(c_i)$ for $x_{i-1} \leq x \leq x_i$, and the work done from x_{i-1} to x_i is approximately $f(c_i)(x_i - x_{i-1})$. The total work done from a to b is approximately $\sum_{i=1}^{n} f(c_i)(x_i - x_{i-1})$. As the lengths of the integrals get smaller, so $\max_i(x_i - x_{i-1}) \longrightarrow 0$, these better and better approximations approach the definite integral as a limit, and that is the work done by $f(x)$ from a to b:

$$W = \int_a^b f(x)\, dx.$$

When setting up such a work problem we can skip the intermediate step of approximating the work over many small intervals. Think of the increment of work done by the force $f(x)$ over the interval of length dx as $f(x)\, dx$, and the total work as the sum (integral) of all these increments.

EXAMPLE 13.1

The force F required to stretch a steel spring is proportional to the distance the spring is stretched, so $F = kx$ where k is a constant. Suppose a two-pound force stretches a spring 4 inches. How much work is done in stretching the spring from 4 inches to 12 inches?

Solution

First, we determine the constant k, and since work is generally measured in foot pounds we measure distance in feet. We are given that 2 lbs corresponds to 4 in ($\frac{1}{3}$ ft), so $2 = \frac{1}{3}k$, and $k = 6$ lbs/ft. Therefore,

$F(x) = 6x$. The work done in stretching from $\frac{1}{3}$ foot to 1 foot is

$$W = \int_{\frac{1}{3}}^{1} 6x\,dx = 3x^2\Big]_{\frac{1}{3}}^{1} = 3 - \frac{1}{3} = \frac{8}{3} \text{ ft–lbs.}$$

EXAMPLE 13.2

How much work is done in pumping the water out of a hemispherical tank of radius 4 ft?

Solution

(See Figure 13.1). We think of the increment of work as the work done in lifting a thin "slice" of water to the top of the tank. The area of a slice x units down from the top is $\pi(4^2 - x^2)$, and the thickness is dx. If the density of water is 62 lbs/ft^3, then the weight of the slice is $62\pi(4^2 - x^2)\,dx$, and the work done lifting it x feet is

$$62\pi(16 - x^2)x\,dx.$$

The work done emptying the tank is

$$\begin{aligned} W &= \int_0^4 62\pi(16 - x^2)x\,dx \\ &= 62\pi \int_0^4 (16x - x^3)\,dx \\ &= 62\pi \left[8x^2 - \frac{1}{4}x^4\right]_0^4 \\ &= 62\pi[128 - 64] \\ &= 62 \cdot 64\pi = 3968\pi \text{ ft lbs.} \end{aligned}$$

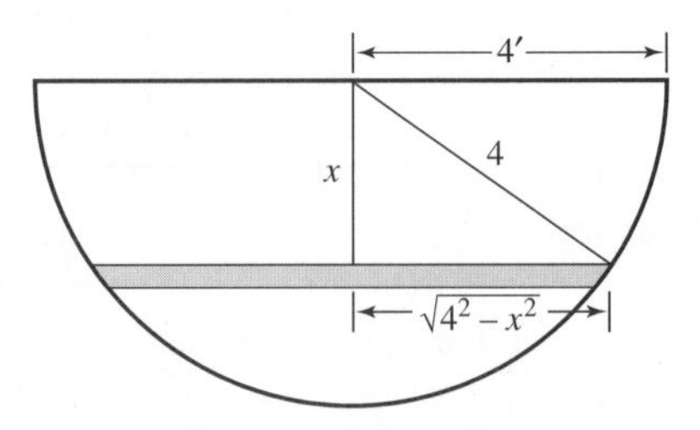

Figure 13.1

Another application of the integral involves calculation of the volume of various solids. Suppose we have a solid whose horizontal cross-section at any given height y is known. In the simplest instance, a rectangular parallelopiped (i.e., a brick), the cross-sectional area A is constant, and the volume V is A times the height h. In general, the cross-sectional area $A(y)$ will depend on the height. The volume of a thin slice at height y is $A(y)\,dy$, where dy is the thickness of the slice. The total volume is the sum of all these incremental volumes:

$$V = \int_0^h A(y)\,dy.$$

EXAMPLE 13.3

Find the volume of a cone of base radius r and height h (Figure 13.2).

Solution

The cross-section at a distance y from the vertex is a circle of radius x, where $\frac{x}{r} = \frac{y}{h}$, or $x = \frac{y}{h}r$. The

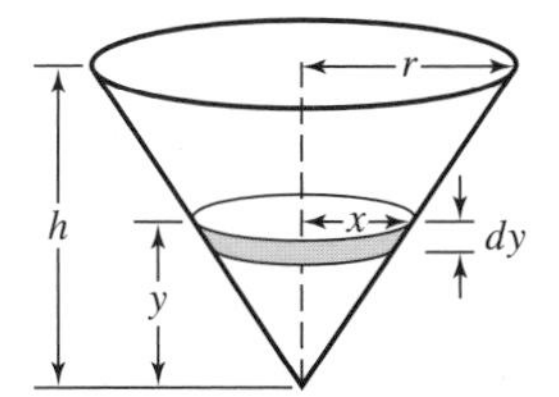

Figure 13.2

cross-sectional area is $A(y) = \pi(\frac{r}{h}y)^2$, and

$$V = \int_0^h \frac{\pi r^2}{h^2} y^2\,dy$$
$$= \frac{\pi r^2}{h^2}\frac{1}{3}y^3\Big]_0^h$$
$$= \frac{1}{3}\pi r^2 h.$$

EXAMPLE 13.4

The base of a solid is a semicircle of radius a, and every cross-section perpendicular to one diameter is a square. Find the volume.

Solution

(Figure 13.3). The cross-section at x has area y^2 where $x^2 + y^2 = a^2$, so the area is $a^2 - x^2$. The thickness is dx so $dV = (a^2 - x^2)\,dx$, and

$$V = \int_{-a}^{a} (a^2 - x^2)\,dx.$$

By the symmetry of the figure, we can write

$$V = 2\int_0^a (a^2 - x^2)\,dx$$
$$= 2\left[a^2x - \frac{1}{3}x^3\right]_0^a$$
$$= 2\left[a^3 - \frac{1}{3}a^3\right] = \frac{4}{3}a^3.$$

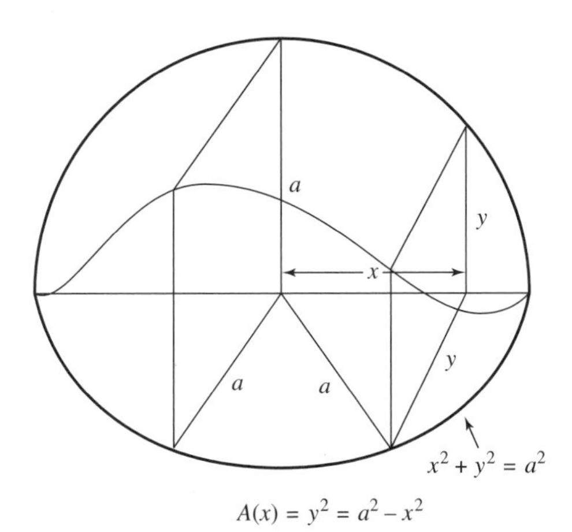

Figure 13.3

If a plane region is rotated about an axis, the volume swept out is called a volume of revolution. The cone of Example 13.3 is the volume obtained by rotating about the y-axis the triangular area between the line $x = \frac{r}{h}y$ and the y-axis, for $0 \le y \le h$. If the area under $y = f(x)$, $a \le x \le b$ is rotated about the x-axis, the volume of revolution will have circular cross-sections of radius $f(x)$ and hence cross-sectional area $\pi f(x)^2$. The increments of volume—discs of thickness dx—will be $\pi f(x)^2\, dx$, so the total volume is

$$V = \int_a^b \pi f(x)^2\, dx.$$

EXAMPLE 13.5

Find the volume of the solid swept out by rotating one arch of the sine curve about the x-axis.

Solution

The sine curve has one arch between 0 and π, so the volume of revolution of this area is (see Figure 13.4)

$$V = \pi \int_0^{\pi} \sin^2 x\, dx.$$

To evaluate the integral, recall the identity

$$\sin^2 x = \frac{1 - \cos 2x}{2}.$$

Hence,

$$\begin{aligned} V &= \pi \int_0^{\pi} \left(\frac{1}{2} - \frac{1}{2}\cos 2x\right) dx \\ &= \pi \left[\frac{1}{2}x - \frac{1}{4}\sin 2x\right]_0^{\pi} \\ &= \pi \left[\left(\frac{\pi}{2} - \frac{1}{4}\sin 2\pi\right) - (0 - \sin 0)\right] \\ &= \frac{\pi^2}{2}. \end{aligned}$$

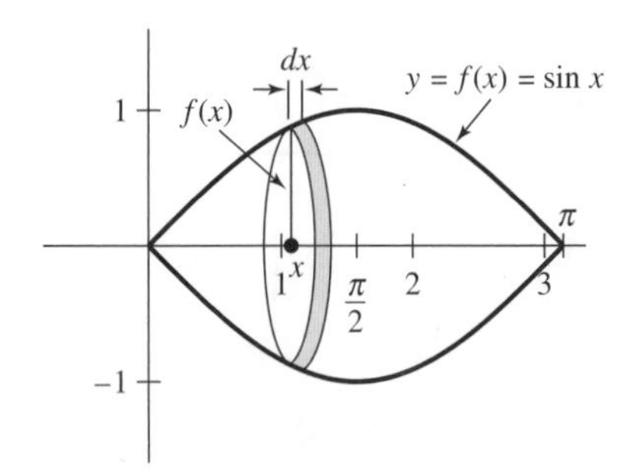

Figure 13.4

As a final application, consider the problem of computing the force exerted on a vertical surface by liquid pressure. Water weighs about 62 pounds per cubic foot, so the pressure at a depth y is $62y$ pounds per square foot. Consider a rectangular fish-viewing window in an aquarium tank. The window is 6 feet long and 4 feet deep, and the top of the window is one foot below the surface of the water. What is the force on the window? Force is pressure times area, so we divide the window up into thin horizontal strips where the depth, and hence the pressure, is constant. At depth y, the pressure is $62y$, and the force on a strip 6 feet long and dy

high is $dF = (62y)(6dy)$. The total force on the window, with y ranging from 1 to 5 feet, is

$$\begin{aligned} F &= \int_1^5 (62)(6)y\,dy \\ &= \frac{62 \cdot 6}{2} y^2 \Big]_1^5 \\ &= 62 \cdot 3 \cdot (25 - 1) = 4464 \text{ lbs.} \end{aligned}$$

PROBLEMS

13.1 Find the work done in stretching a spring 2 feet, if a force of 3 pounds is required to stretch the spring 6 inches.

13.2 A 20-pound weight is hung from a spring, and the spring stretches 8 inches. How much work is done in pulling the weight down an additional foot?

13.3 A leaking bucket of water is lifted 50 feet. The bucket weighs 60 pounds initially and loses weight uniformly until it weighs 30 pounds at the top. How much work was done?

13.4 How much work is done in pumping the water out the top of a cylindrical tank whose radius is 4 feet and whose depth is 6 feet? *Hint*: Consider the increment of work done in lifting a "slice" of water y feet from the top and dy thick. The slice weighs $62 \cdot \pi \cdot 4^2 dy$, and the work done lifting this slice up y units is its weight times y.

13.5 Find the work done in pumping the top 2 feet of water out of a hemispherical bowl of radius 5 feet. *Hint*: The slice at depth y feet has radius $x = \sqrt{5^2 - y^2}$.

13.6 An irregular solid has horizontal cross-sectional area at height y equal to $A(y) = \sqrt{9 - y}$, for $0 \le y \le 9$. What is its volume?

13.7 In Example 13.4 suppose the solid again has a base that is a semicircle of radius a, but now the vertical cross-sections are quarter circles. Find the volume.

13.8 Find the volume of the sphere obtained by rotating about the x-axis the area under the curve $y = \sqrt{a^2 - x^2}$, $-a \le x \le a$.

13.9 What is the volume obtained by rotating the area under the parabola $y = x^2$, for $0 \le x \le 1$, about the x-axis?

13.10 Rotate the area of Problem 13.9 about the y-axis. What is the volume? *Hint*: The slices for fixed y are washers with thickness dy, outer radius 1, inner radius $\sqrt{y}$.

13.11 Rotate about the y-axis the area between the curve $y = \sin^{-1} x$ and the y-axis, for $0 \le y \le \frac{\pi}{2}$. What is the volume?

13.12 Find the volume of the ellipsoid obtained by rotating the ellipse $\frac{x^2}{a^2} + \frac{y^2}{b^2} = 1$ about the x-axis. Then find the volume of rotation about the y-axis.

13.13 A body moving with constant velocity v goes a distance $v(t_2 - t_1)$ between times t_1 and t_2. If $v(t)$ is a varying velocity, the distance is $\int_{t_1}^{t_2} v(t)\,dt$. Show that if a body has constant acceleration a, so $v = at$, then the distance it travels in t seconds is $\frac{1}{2}at^2$. (If a is the acceleration of gravity, 32 ft/sec^2, then this is the falling body again.)

13.14 Suppose a car has a constant acceleration and goes from 0 to 60 mph in 6 seconds. How far does the car go in these 6 seconds? *Hint*: 60 mph is 88 ft/sec, so $v = \frac{88}{6}t$ ft/sec at time t.

13.15 The shallow end of a swimming pool is a vertical rectangle, 4 feet deep and 20 feet across. What is the force on this surface exerted by the water when the pool is full?

13.16 A 15-foot chain weighing 2 lbs/ft lies coiled on the ground. A line of negligible weight is attached to one end and used to lift the chain straight up until the bottom just clears the ground. How much work was done?

13.17 A 10-foot chain weighing $\frac{1}{2}$ lb per foot hangs from a roof. How much work is done in pulling the chain up onto the roof?

14

Parametric Equations

So far we have described curves with equations of the form $y = f(x)$ or $F(x, y) = 0$. To describe the path of a moving object, it is frequently more convenient and more relevant to determine the individual coordinates as functions of time t; thus,

$$x = g(t), \quad y = f(t). \tag{14.1}$$

Equations (14.1) are called **parametric equations** of a curve, and the variable t is called the **parameter**.

EXAMPLE 14.1

A cannon fires a projectile with muzzle velocity v at an elevation θ from the horizontal. Describe the path of the projectile.

Solution

(See Figure 14.1.) The horizontal component of the velocity is $v \cos\theta$, and that is the constant speed in the x-direction. Therefore, with the cannon at the origin,

$$x = (v \cos\theta)t.$$

The initial vertical component of the velocity is $v \sin\theta$, so as we have seen earlier, the motion in the vertical direction is given by

$$y = (v \sin\theta)t - 16t^2,$$

where the term $-16t^2$ is due to the downward acceleration of gravity. The two equations

$$\begin{aligned} x &= (v \cos\theta)t \\ y &= (v \sin\theta)t - 16t^2 \end{aligned} \tag{14.2}$$

are parametric equations of the path of the projectile.

The Cartesian equation that corresponds to a given pair of parametric equations can be found by eliminating the parameter to get an equation in x and y. For example, in (14.2) we could solve the first equation for t and substitute that expression in the second equation. This

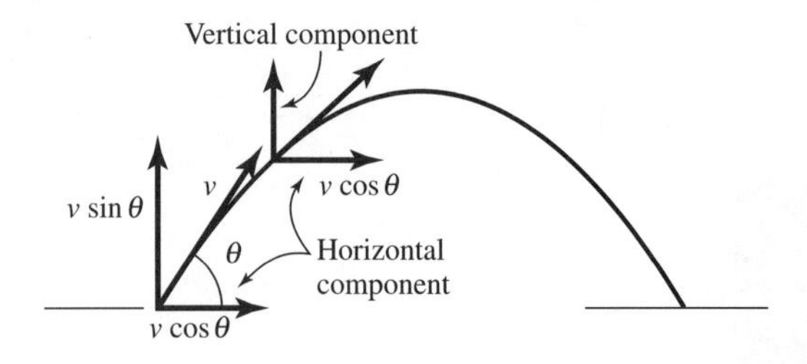

Figure 14.1

gives

$$t = x/v\cos\theta,$$
$$y = (v\sin\theta)\left(\frac{x}{v\cos\theta}\right) - 16\left(\frac{x}{v\cos\theta}\right)^2$$
$$= (\tan\theta)x - \frac{16}{v^2\cos^2\theta}x^2.$$

Since y is a quadratic function of x (remember that θ and v are given constants), the path is a parabola.

In Example 14.1, the parameter is time, which is most appropriate when the curve is the path of a moving object. Curves can also be described in terms of a parameter that has purely geometric significance. A simple illustration is the curve

$$\begin{aligned} x &= \cos\theta \\ y &= \sin\theta \end{aligned} \tag{14.3}$$

for $0 \le \theta \le \pi$. If the parameter were unrestricted, then equations (14.3) would be the parametric equations of the unit circle, and the parameter θ would be the distance you would have to go around the circle from $(1, 0)$ to get to (x, y). Since in (14.3) θ is restricted to $[0, \pi]$, (14.3) represents just the top half of the unit circle.

A parametrically defined curve may consist of only a part of the corresponding Cartesian curve that results when the parameter is eliminated. Consider the curve

$$x = \sin\theta, \quad y = 1 - \sin^2\theta. \tag{14.4}$$

The corresponding Cartesian equation is

$$y = 1 - x^2. \tag{14.5}$$

Notice, however, that in equation (14.4) x takes only values between -1 and 1, and y takes only values between 0 and 1. The parametric curve (14.5) therefore consists only of the part of the parabola (14.5) for $-1 \le x \le 1$.

To find the slope of a parametric curve $x = x(t)$, $y = y(t)$, let Δx and Δy be the corresponding changes in x and y, both of which result from a change Δt in t. Then

$$\begin{aligned} \frac{dy}{dx} &= \lim_{\Delta t\to 0}\frac{\Delta y}{\Delta x} \\ &= \lim_{\Delta t\to 0}\frac{\Delta y/\Delta t}{\Delta x/\Delta t} \\ &= \frac{dy}{dt}\Big/\frac{dx}{dt}. \end{aligned} \tag{14.6}$$

For example, the slope of the parabola (14.4) at the point corresponding to a given θ is

$$\frac{dy}{dx} = \frac{\frac{dy}{d\theta}}{\frac{dx}{d\theta}} = \frac{-2\sin\theta\cos\theta}{\cos\theta} = -2\sin\theta.$$

When $\theta = \frac{\pi}{2}$, so the point is $(1, 0)$,

$$\frac{dy}{dx} = -2\sin\frac{\pi}{2} = -2.$$

When x and y are given in terms of a parameter t, then the calculation of the slope gives $\frac{dy}{dx}$ as a function of t. To find $\frac{d^2y}{dx^2}$ we simply use (14.6) again to find the derivative of $\frac{dy}{dx}$ with respect to x in this situation where both x and $\frac{dy}{dx}$ are functions of t. That is

$$\frac{d^2y}{dx^2} = \frac{\frac{d}{dt}\left(\frac{dy}{dx}\right)}{\frac{dx}{dt}}. \tag{14.7}$$

EXAMPLE 14.2

Find the tangent line to $y = t^2 + 1$, $x = 3t - 2$ at $t = 1$, and tell whether the curve lies over or under the tangent line near $(x(1), y(1))$.

Solution

First calculate $\frac{dy}{dx}$:

$$\frac{dy}{dx} = \frac{\frac{dy}{dt}}{\frac{dx}{dt}} = \frac{2t}{3}.$$

At $t = 1$, $x = 1$, $y = 2$, and $\frac{dy}{dx} = \frac{2}{3}$. Hence, the tangent line at $(1, 2)$ has the equation

$$y - 2 = \frac{2}{3}(x - 1).$$

Recall that if $\frac{d^2y}{dx^2}$ is positive, the first derivative is increasing, and the curve is **concave up**. A curve that is concave up will lie above its tangent lines, and a curve that is concave down ($\frac{d^2y}{dx^2} < 0$) will lie below its tangent lines. We use (14.7) to calculate $\frac{d^2y}{dx^2}$:

$$\begin{aligned}\frac{d^2y}{dx^2} &= \frac{\frac{d}{dt}\left(\frac{dy}{dx}\right)}{\frac{dx}{dt}} \\ &= \frac{\frac{d}{dt}\left(\frac{2t}{3}\right)}{3} = \frac{2}{9}.\end{aligned}$$

The curve has a constant positive second derivative and so is concave up everywhere, and the curve lies over every tangent line. To change to Cartesian coordinates, solve for t, $t = \frac{1}{3}(x + 2)$, and substitute in the formula for y:

$$\begin{aligned}y &= \left[\frac{1}{3}(x + 2)\right]^2 + 1 \\ &= \frac{1}{9}(x + 2)^2 + 1.\end{aligned}$$

The curve is an upward-pointing parabola.

Let $f(t)$ and $g(t)$ be any two differentiable functions on an interval $[a, b]$, and consider the curve

$$x = g(t), \quad y = f(t), \quad a \le t \le b.$$

Suppose that $g'(t) > 0$ on $[a, b]$ so that x increases and the curve moves from left to right as t goes from a to b. By the ordinary Mean Value Theorem, there must be some point $(g(t_0), f(t_0))$ on the curve between $(g(a), f(a))$ and $(g(b), f(b))$ where the slope of the curve is the same as the slope of the secant line; namely, $(f(b) - f(a))/(g(b) - g(a))$. Since the slope of the curve is given by

$$\frac{dy}{dx} = \frac{\frac{dy}{dt}}{\frac{dx}{dt}} = \frac{f'(t)}{g'(t)},$$

this says that there must be some t_0 in (a, b) such that

$$\frac{f(b) - f(a)}{g(b) - g(a)} = \frac{f'(t_0)}{g'(t_0)}. \tag{14.8}$$

The statement above is called **Cauchy's Mean Value Theorem**. The two forms of the Mean Value Theorem look different, since Cauchy's version (14.8) involves two functions, but the geometric content is identical. A curve cannot get smoothly from P to Q without somewhere pointing in the same direction as the line segment joining P and Q.

The length of a curve is defined to be the limit of the lengths of polygonal paths joining consecutive points along the curve. If the curve is given by parametric equations $x = x(t)$, $y = y(t)$ for $a \le t \le b$, then for each partition $a = t_0 < t_1 < t_2 < \cdots < t_n = b$ of $[a, b]$, we get a polygonal path whose length is

$$\sum_{i=1}^{n} \sqrt{\Delta x_i^2 + \Delta y_i^2},$$

where $\Delta x_i = x(t_{i+1}) - x(t_i)$ and $\Delta y_i = y(t_{i+1}) - y(t_i)$. We can write the length of the polygon as

$$\sum_{i=1}^{n} \sqrt{\Delta x_i^2 + \Delta y_i^2} = \sum_{i=1}^{n} \sqrt{\left(\frac{\Delta x_i}{\Delta t_i}\right)^2 + \left(\frac{\Delta y_i}{\Delta t_i}\right)^2}\, \Delta t_i. \tag{14.9}$$

As the partition becomes finer and finer, with $\max \Delta t_i \longrightarrow 0$, the sums in (14.9) approach the following integral, which is the arc length:

$$s = \int_a^b \sqrt{\left(\frac{dx}{dt}\right)^2 + \left(\frac{dy}{dt}\right)^2}\, dt. \tag{14.10}$$

If the curve is given in the form $y = f(x)$ for $a \le x \le b$, this is the same as the parametric form

$$x = t, \quad y = f(t), \quad a \le t \le b,$$

and the integral (14.10) becomes

$$\begin{aligned} s &= \int_a^b \sqrt{1 + \left(\frac{dy}{dx}\right)^2}\, dx \\ &= \int_a^b \sqrt{1 + (f'(x))^2}\, dx. \end{aligned} \tag{14.11}$$

EXAMPLE 14.3

Find the length of the curve $x = 2t$, $y = \frac{4}{3}t^{3/2}$, $0 \le t \le 1$.

Solution

Since $\frac{dx}{dt} = 2$ and $\frac{dy}{dt} = 2t^{\frac{1}{2}}$,

$$\begin{aligned} s &= \int_0^1 \sqrt{4 + 4t}\, dt \\ &= 2\int_0^1 \sqrt{1 + t}\, dt \\ &= \frac{4}{3}(1 + t)^{\frac{3}{2}}\Big]_0^1 = \frac{4}{3}(2^{\frac{3}{2}} - 1). \end{aligned}$$

PROBLEMS

Graph the following curves. Notice that both x and y have a specific finite range of values for all these curves. Indicate these ranges clearly on your graph.

14.1 $x = t, \quad y = t^2, \quad 0 \le t \le 1$

14.2 $x = t^2, \quad y = 2t - 1, \quad 0 \le t \le 1$

14.3 $x = 2\cos t, \quad y = \sin t, \quad 0 \le t \le \pi$

14.4 $x = \frac{1}{t}, \quad y = t, \quad 1 \le t \le 2$

Write the Cartesian equations of the following curves, and graph them.

14.5 $x = \sin^2 t + 1, \quad y = \cos t, \quad -\infty < t < \infty$

14.6 $x = t^2, \quad y = 2t^4 - 1, \quad -\infty < t < \infty$

14.7 $x = 1 - e^t, \quad y = e^{2t}, \quad -\infty < t < \infty$

14.8 $x = \sin^2 t, \quad y = \cos^2 t, \quad 0 \le t \le \dfrac{\pi}{2}$

Find the equation of the tangent line at the indicated point. Does the curve lie under or over the tangent line near this point?

14.9 $x = 2t^3 - 1, \quad y = t^2, \quad t = 2$

14.10 $x = e^t, \quad y = \cos t, \quad t = 0$

14.11 $x = t^2 - 1, \quad y = t^2 + 1, \quad t = 113$

14.12 $x = e^t, \quad y = e^{2t}, \quad t = 0$

14.13 A ball is thrown at an angle of 45° with the ground and an initial velocity of 64 ft/sec. Find how high it goes by finding when $\frac{dy}{dt} = 0$.

14.14 If a body is projected upward at an angle θ, we have seen that its path is given by $x = (v\cos\theta)t$, $y = (v\sin\theta)t - 16t^2$, where v is the initial velocity. At what time does the body reach maximum height, and when does it hit the ground? At what angle does the curve hit the ground? What is the slope of the curve when the body hits the ground?

Find the lengths of the following curves.

14.15 $x = a\cos t, \quad y = a\sin t, \quad 0 \le t \le 2\pi$

14.16 $y = \dfrac{8}{3}x^{\frac{3}{2}}, \quad 0 \le x \le 5$

14.17 $x = \sqrt{1+t^2}, \quad y = \log(t + \sqrt{1+t^2}), \quad 0 \le t \le 5$

14.18 $x = \tan^{-1} t, \quad y = \frac{1}{2}\log(1+t^2), \quad 0 \le t \le 1$

14.19 $x = e^t\cos t, \quad y = e^t\sin t, \quad 0 \le t \le \pi$

14.20 $y = \cosh x, \quad 0 \le x \le x_0$ (See Problem 7.28, Chapter 7.)

14.21 Find the parametric equations of the hyperbola $\frac{x^2}{a^2} - \frac{y^2}{b^2} = 1$ using the parameter θ defined by $x = a\sec\theta$. *Hint*: $\tan^2\theta + 1 = \sec^2\theta$.

14.22 Find the parametric equations of the parabola $ay = x^2$ using as parameter the slope of the line from $(0, 0)$ to (x, y).

14.23 If a circle rolls along the x-axis, the point P on the circle that starts at $(0, 0)$ traces out a **cycloid**. Show that the parametric equations are $x = a\theta - a\sin\theta$, $y = a - a\cos\theta$, where a is the radius of the circle and θ is the angle through which the radius to P has turned.

14.24 Find the parametric equations of the ellipse $\frac{x^2}{a^2} + \frac{y^2}{b^2} = 1$ using as parameter the angle θ defined by $x = a\cos\theta$.

14.25 A rod AB moves with its end A on the y-axis and its end B on the x-axis. Find the parametric equations of the curve traced out by the point P on the rod which is a units from A and b units from B. *Hint*: Use as parameter the angle θ the rod makes with the x-axis.

15

Change of Variable

We have already seen that most integration problems involve the chain rule in reverse, and the u-substitution is a useful way to keep track of the constants in these problems. For example, to integrate

$$\int x^2(x^3+1)^4\,dx$$

we first recognize that x^2 is the essential part of the derivative of x^3+1, and we let

$$u = x^3+1, \quad du = 3x^2\,dx, \quad x^2\,dx = \frac{1}{3}\,du.$$

This gives

$$\int x^2(x^3+1)^4\,dx = \int \frac{1}{3}u^4\,du = \frac{1}{15}u^5 = \frac{1}{15}(x^3+1)^5. \tag{15.1}$$

For definite integrals we can carry this technique one step further and incorporate the limits of integration in the change of variable. For example, using the final answer in (15.1) we can evaluate the following definite integral:

$$\int_{-1}^{1} x^2(x^3+1)^4\,dx = \frac{1}{15}(x^3+1)^5\Big]_{-1}^{1} = \frac{2^5}{15}.$$

Here we ignored the intermediate steps involving u and used just the final antiderivative $(\frac{1}{15})(x^3+1)^5$. Instead of this approach, notice that if $u = (x^3+1)$, then $u = 0$ when $x = -1$ and $u = 2$ when $x = 1$. Therefore, we can make the u-substitution with the new limits:

$$\int_{-1}^{1} x^2(x^3+1)^4\,dx = \int_0^2 \frac{1}{3}u^4\,du = \frac{1}{15}u^5\Big]_0^2 = \frac{1}{15}\cdot 2^5. \tag{15.2}$$

The general rule for changing the variable in a definite integral is the following formula:

$$\int_a^b f(u(x))u'(x)\,dx = \int_{u(a)}^{u(b)} f(u)\,du. \tag{15.3}$$

In Equation (15.2),

$$u = x^3 + 1, \quad f(u(x)) = \frac{1}{3}(x^3+1)^4,$$

$$u'(x)\,dx = 3x^2\,dx, \quad f(u)\,du = \frac{1}{3}u^4\,du,$$

$$u(-1) = 0, \quad u(1) = 2.$$

To verify the change of variable equation (15.3), we let $F(x)$ be any antiderivative of $f(x)$, so $F'(x) = f(x)$. It follows that

$$\int_{u(a)}^{u(b)} f(u)\,du = F(u)\Big]_{u(a)}^{u(b)} = F(u(b)) - F(u(a)). \tag{15.4}$$

From the chain rule we also have

$$\frac{d}{dx}F(u(x)) = F'(u(x))u'(x) = f(u(x))u'(x),$$

so $F(u(x))$ is an antiderivative of the integrand on the left of (15.3). Therefore,

$$\int_a^b f(u(x))u'(x)\,dx = F(u(x))\Big]_a^b = F(u(b)) - F(u(a)). \tag{15.5}$$

Comparison of (15.4) and (15.5) verifies the change of variable equation (15.3).

Here are some more examples.

EXAMPLE 15.1

$$\int_2^3 \sqrt{3x-5}\,dx.$$

Let $u = 3x - 5$, $du = 3\,dx$, $dx = \frac{1}{3}\,du$. When $x = 2$, $u = 1$ and when $x = 3$, $u = 4$. Therefore,

$$\int_2^3 \sqrt{3x-5}\,dx = \int_1^4 \frac{1}{3}\sqrt{u}\,du = \frac{1}{3}\cdot\frac{2}{3}u^{\frac{3}{2}}\Big]_1^4 = \frac{2}{9}(8-1) = \frac{14}{9}.$$

EXAMPLE 15.2

$$\int_0^{\frac{\pi}{2}} \sin^2 x \cos x dx.$$

Let $u = \sin x$, $du = \cos x\,dx$. When $x = 0$, $u = 0$ and when $x = \frac{\pi}{2}$, $u = 1$. Therefore,

$$\int_0^{\frac{\pi}{2}} \sin^2 x \cos x\,dx = \int_0^1 u^2\,du = \frac{1}{3}u^3\Big]_0^1 = \frac{1}{3}.$$

EXAMPLE 15.3

$$\int_1^5 \frac{\log x}{x}\,dx.$$

Let $u = \log x$, $du = \frac{1}{x}\,dx$. When $x = 1$, $u = 0$ and when $x = 5$, $u = \log 5$. Therefore,

$$\int_1^5 \frac{\log x}{x}\,dx = \int_0^{\log 5} u\,du = \frac{1}{2}u^2\Big]_0^{\log 5} = \frac{1}{2}(\log 5)^2.$$

EXAMPLE 15.4

$$\int_{-1}^0 xe^{x^2}\,dx.$$

Let $u = x^2$, $du = 2x\,dx$, $x\,dx = \frac{1}{2}\,du$. When $x = -1$, $u = 1$ and when $x = 0$, $u = 0$. Therefore,

$$\int_{-1}^0 xe^{x^2}\,dx = \int_1^0 \frac{1}{2}e^u\,du = \frac{1}{2}e^u\Big]_1^0 = \frac{1}{2}(1-e).$$

The answer is negative, which we expect since the first integrand xe^{x^2} is negative on $[-1, 0]$. The integrand after the substitution, $\frac{1}{2}e^u$, is positive, but the integration is from right to left (from 1 to 0), so the result is negative.

We have already used the important differentiation–integration formulas

$$\frac{d}{dx}\tan^{-1}x = \frac{1}{1+x^2}; \quad \int \frac{dx}{1+x^2} = \tan^{-1}x;$$

$$\frac{d}{dx}\sin^{-1}x = \frac{1}{\sqrt{1-x^2}}; \quad \int \frac{dx}{\sqrt{1-x^2}} = \sin^{-1}x.$$

For integration problems, the following more general forms are convenient:

$$\int \frac{dx}{a^2+x^2} = \frac{1}{a}\tan^{-1}\frac{x}{a}; \tag{15.6}$$

$$\int \frac{dx}{\sqrt{a^2-x^2}} = \sin^{-1}\frac{x}{a}. \tag{15.7}$$

EXAMPLE 15.5

$\int_0^{\sqrt{5}} \frac{dx}{5+x^2}$.

This is (15.6) with $a^2 = 5$, $a = \sqrt{5}$, so

$$\int_0^{\sqrt{5}} \frac{dx}{5+x^2} = \frac{1}{\sqrt{5}}\tan^{-1}\frac{x}{\sqrt{5}}\Big]_0^{\sqrt{5}}$$

$$= \frac{1}{\sqrt{5}}\left[\frac{\pi}{4} - 0\right] = \pi/4\sqrt{5}.$$

EXAMPLE 15.6

$\int_0^1 \frac{dx}{\sqrt{4-2x^2}}$.

Let $u = \sqrt{2}x$ so $u^2 = 2x^2$ and $dx = \frac{1}{\sqrt{2}}\,du$. When $x = 0$, $u = 0$ and when $x = 1$, $u = \sqrt{2}$. Therefore, using (15.7),

$$\int_0^1 \frac{dx}{\sqrt{4-2x^2}} = \int_0^{\sqrt{2}} \frac{\frac{1}{\sqrt{2}}\,du}{\sqrt{4-u^2}}$$

$$= \frac{1}{\sqrt{2}}\sin^{-1}\frac{u}{2}\Big]_0^{\sqrt{2}}$$

$$= \frac{1}{\sqrt{2}}\left[\sin^{-1}\frac{\sqrt{2}}{2} - 0\right]$$

$$= \frac{1}{\sqrt{2}}\frac{\pi}{4}.$$

PROBLEMS

Make a change of variable to evaluate the following integrals.

15.1 $\int_0^1 x(x^2+3)^5\,dx$

15.2 $\int_1^2 \frac{dx}{3+2x}$

15.3 $\int_1^5 \sqrt{2x-1}\,dx$

15.4 $\int_{-1}^{1} x\sqrt{1-x^2}\,dx$

15.5 $\int_{0}^{\frac{\pi}{4}} \cos x \sin x\,dx$

15.6 $\int_{0}^{\frac{\pi}{2}} \sqrt{\sin x}\cos x\,dx$

15.7 $\int_{1}^{2} \frac{\log 2x}{x}\,dx$

15.8 $\int_{e}^{e^2} \frac{(\log x)^3}{x}\,dx$

15.9 $\int_{0}^{4} \frac{x^2+1}{x+1}\,dx$ (Divide first.)

15.10 $\int_{0}^{1} 3xe^{x^2}\,dx$

15.11 $\int_{1}^{4} \frac{e^{\sqrt{x}}}{\sqrt{x}}\,dx$

15.12 $\int_{0}^{2} \frac{dx}{4+x^2}$

15.13 $\int_{0}^{1} \frac{dx}{1+4x^2}$

15.14 $\int_{0}^{3} \frac{x dx}{1+x^2}$

15.15 $\int_{0}^{1} \frac{e^x dx}{1+e^{2x}}$

15.16 $\int_{0}^{1} \frac{dx}{\sqrt{16-x^2}}$

15.17 $\int_{0}^{1} \frac{dx}{\sqrt{5-3x}}$

15.18 $\int_{0}^{\pi} \sin 2x\,dx$

15.19 $\int_{1}^{2} \frac{dx}{9+4x^2}$

15.20 $\int_{0}^{1} \frac{dx}{\sqrt{16-9x^2}}$

15.21 $\int_{1}^{3} \frac{dx}{4+25x^2}$

15.22 $\int_{0}^{7} \frac{2x+1}{x^2+x+5}$

15.23 $\int_{1}^{2} x(x-1)^{10}dx$ Let $u = x-1$.

15.24 $\int_{2}^{7} x\sqrt{x+2}dx$ Let $u = x+2$.

15.25 $\int_{0}^{1} \frac{x+2}{\sqrt{x+1}}\,dx$

15.26 $\int_{1}^{2} x^2(x-1)^{\frac{1}{3}}\,dx$

15.27 $\int_{0}^{1} \frac{x^2}{\sqrt{x+2}}\,dx$

16

Integrating Rational Functions

A **rational function** is an expression $\frac{P(x)}{Q(x)}$, where $P(x)$ and $Q(x)$ are polynomials. There is a standard algorithm for finding the antiderivative of any rational function in which you can completely factor the denominator. In most cases, the method requires a ridiculous amount of algebra and is not practical. However, if $Q(x)$ is linear or quadratic, or a product of distinct linear factors, the method is quite straightforward.

The first thing to notice when integrating *any* rational function $\frac{P(x)}{Q(x)}$ is that if $P(x)$ has degree greater than or equal to the degree of $Q(x)$, you can divide and get

$$\frac{P(x)}{Q(x)} = S(x) + \frac{R(x)}{Q(x)},$$

where $S(x)$ is a polynomial, and $R(x)$ has degree strictly smaller than that of $Q(x)$. The polynomial $S(x)$ is easy to integrate, so we only have to consider rational functions $\frac{R(x)}{Q(x)}$ in which the numerator has smaller degree than the denominator. **IMPORTANT**: *First divide if you can.*

Now we consider only cases where the numerator has smaller degree. If $Q(x)$ is linear, the integration is straightforward.

$$\int \frac{k}{x-a}\,dx = k \log|x-a|.$$

Now suppose $Q(x)$ is quadratic. We can always factor out the coefficient of x^2, so we assume $Q(x)$ has the form $Q(x) = x^2 + Bx + C$. There are three cases to consider:

(i) $Q(x) = (x-a)^2$

(ii) $Q(x) = (x-a)(x-b)$ with $a \neq b$

(iii) $Q(x) = x^2 + Bx + C$ with $B^2 - 4C < 0$, so $Q(x)$ does not factor.

We illustrate each case with a specific example which shows the technique.

Case (i). $\int \frac{2x+4}{(x-3)^2}\,dx.$

We substitute $u = x - 3$, $x = u + 3$, and $du = dx$.

$$\begin{aligned}\int \frac{2x+4}{(x-3)^2}\,dx &= \int \frac{2(u+3)+4}{u^2}\,dx \\ &= \int \frac{2u}{u^2}\,du + \int \frac{10}{u^2}\,du \\ &= 2\log|u| - 10u^{-1} \\ &= 2\log|x-3| - 10(x-3)^{-1}.\end{aligned}$$

Case (ii). $\int \frac{2x+7}{(x-1)(x+2)}\,dx.$

Since the denominator has distinct factors, there will be constants A and B such that

$$\begin{aligned}\frac{2x+7}{(x-1)(x+2)} &= \frac{A}{x-1} + \frac{B}{x+2} \\ &= \frac{A(x+2)+B(x-1)}{(x-1)(x+2)}.\end{aligned} \tag{16.1}$$

For (16.1) to be an identity, the numerators must be identical:

$$2x + 7 = A(x+2) + B(x-1). \tag{16.2}$$

Let $x = -2$ and this becomes $3 = -3B$, $B = -1$. Let $x = 1$ and (16.2) becomes $9 = 3A$, $A = 3$. Therefore, the integrand has the following **partial fractions decomposition:**

$$\frac{2x+7}{(x-1)(x+2)} = \frac{3}{x-1} + \frac{-1}{x+2}.$$

Now we can integrate

$$\begin{aligned}\int \frac{2x+7}{(x-1)(x+2)}\,dx &= \int \frac{3}{x-1}\,dx - \int \frac{dx}{x+2} \\ &= 3\log|x-1| - \log|x+2|.\end{aligned}$$

Case (iii). $Q(x) = x^2 + Bx + C$ with $B^2 - 4C < 0$.

In this case, $Q(x)$ can always be written in the form $(x+a)^2 + b^2$, and then the substitution $u = x + a$, $x = u - a$, $du = dx$ works. For example:

$$\int \frac{3x+7}{x^2+2x+5}\,dx.$$

First complete the square in $x^2 + 2x + 5$ to get the form $(x+a)^2 + b^2$:

$$\begin{aligned}x^2 + 2x + 5 &= x^2 + 2x + 1 + 4 \\ &= (x+1)^2 + 4.\end{aligned}$$

Now let $u = x + 1$, $x = u - 1$, $du = dx$:

$$\begin{aligned}
\int \frac{3x+7}{x^2+2x+5}\,dx &= \int \frac{3x+7}{(x+1)^2+4}\,dx \\
&= \int \frac{3(u-1)+7}{u^2+4}\,du \\
&= \int \frac{3u+4}{u^2+4}\,du \\
&= \frac{3}{2}\int \frac{2u\,du}{u^2+4} + 4\int \frac{du}{u^2+4} \\
&= \frac{3}{2}\log|u^2+4| + 2\tan^{-1}\frac{u}{2} \\
&= \frac{3}{2}\log|x^2+2x+5| + 2\tan^{-1}\left(\frac{x+1}{2}\right).
\end{aligned}$$

The partial fractions technique we used for quadratic denominators of the form $(x-a)(x-b)$ will actually work for any rational function $\frac{R(x)}{Q(x)}$, where $R(x)$ has degree less than that of $Q(x)$, and $Q(x)$ can be completely factored into linear and unfactorable quadratic factors. For example, if $Q(x) = (x-a)(x-b)(x-c)$, with a, b, c distinct, then there are numbers A, B, C such that

$$\frac{R(x)}{(x-a)(x-b)(x-c)} = \frac{A}{x-a} + \frac{B}{x-b} + \frac{C}{x-c}.$$

If $Q(x)$ has repeated factors, then these factors must occur in the partial fractions. For example,

$$\frac{R(x)}{(x-a)^3(x-b)^2} = \frac{A}{x-a} + \frac{B}{(x-a)^2} + \frac{C}{(x-a)^3} + \frac{D}{x-b} + \frac{E}{(x-b)^2}.$$

If $Q(x)$ has both linear and nonfactorable quadratic factors, then the decomposition looks like the following.

$$\frac{R(x)}{(x-1)(x^2+4)} = \frac{A}{x-1} + \frac{Bx+C}{(x^2+4)}, \tag{16.3}$$

$$\frac{R(x)}{(x-1)(x-2)^2(x^2+4)^2} = \frac{A}{x-1} + \frac{B}{x-2} + \frac{C}{(x-2)^2} + \frac{Dx+E}{x^2+4} + \frac{Fx+G}{(x^2+4)^2}. \tag{16.4}$$

We know how to integrate all the terms on the right above except $\frac{G}{(x^2+4)^2}$, and we will see how to do this by trigonometric substitution in Chapter 19. The following example illustrates how to find the constants A, B, C, and so on.

EXAMPLE 16.1

$$\int \frac{5x^2-3x+13}{(x-1)(x^2+4)}\,dx.$$

The partial fractions decomposition is (16.4):

$$\frac{5x^2 - 3x + 13}{(x-1)(x^2+4)} = \frac{A}{x-1} + \frac{Bx + C}{x^2+4}$$
$$= \frac{(A+B)x^2 + (C-B)x + (4A - C)}{(x-1)(x^2+4)}.$$

For this to be an identity we must have

$$A + B = 5$$
$$C - B = -3$$
$$4A - C = 13.$$

Adding the corresponding sides of the first two equations gives

$$A + C = 2$$
$$4A - C = 13.$$

Adding again, we find $5A = 15$, $A = 3$; then from the first equation, $B = 2$, and from the second equation $C = -1$. Thus,

$$\int \frac{5x^2 - 3x + 13}{(x-1)(x^2+4)}\, dx = \int \frac{3}{x-1}\, dx + \int \frac{2x-1}{x^2+4}\, dx$$
$$= 3 \log |x-1| + \log |x^2+4| - \frac{1}{2} \tan^{-1} \frac{x}{2}.$$

PROBLEMS

16.1 $\int \frac{3}{x+2}\, dx$

16.2 $\int \frac{1}{(x-3)^3}\, dx$

16.3 $\int \frac{3}{4x+1}\, dx$

16.4 $\int \frac{x+1}{x-1}\, dx$

16.5 $\int \frac{x^2+x+1}{x}\, dx$

16.6 $\int \frac{x^2}{x+4}\, dx$

16.7 $\int \frac{2}{(x-5)(x-2)}\, dx$

16.8 $\int \frac{1}{(x+1)(x-2)}\, dx$

16.9 $\int \frac{x\, dx}{(x-1)(x-2)}$

16.10 $\int \frac{2}{x^2-1}\, dx$

16.11 $\int \frac{dx}{4-x^2}$

16.12 $\displaystyle\int \frac{dx}{x^2+2x+2}$

16.13 $\displaystyle\int \frac{dx}{x^2+4x+5}$

16.14 $\displaystyle\int \frac{3x}{x^2+1}\,dx$

16.15 $\displaystyle\int \frac{2x+1}{x^2+9}\,dx$

16.16 $\displaystyle\int \frac{x+1}{x^2+4}\,dx$

16.17 $\displaystyle\int \frac{4x}{x^2+6x+10}\,dx$

16.18 $\displaystyle\int \frac{dx}{x^2+3x+2}$

16.19 $\displaystyle\int \frac{3x-6}{x^2+4x+8}\,dx$

16.20 $\displaystyle\int \frac{5x^2+2x+3}{(x+1)(x^2+1)}\,dx$

16.21 $\displaystyle\int \frac{6x^2+11x+8}{x(x^2+2x+2)}\,dx$

16.22 $\displaystyle\int \frac{2x^2-10x+2}{(x-1)(x+2)(x-3)}\,dx$ *Hint:*

$$\frac{2x^2-10x+2}{(x-1)(x+2)(x-3)} = \frac{A}{x-1} + \frac{B}{x+2} + \frac{C}{x-3}$$

is equivalent to

$$2x^2-10x+2 = A(x+2)(x-3) + B(x-1)(x-3) + C(x-1)(x+2).$$

Let $x=1$ to find A, $x=-2$ to find B, and $x=3$ to find C.

16.23 $\displaystyle\int \frac{2x^2+7x+4}{x(x+1)(x+2)}\,dx$

16.24 (i) Find A, B, C so that $\frac{4}{x(x^2+4)} = \frac{A}{x} + \frac{Bx+C}{x^2+4}$.
(ii) Integrate $\int \frac{4}{x(x^2+4)}\,dx$.

17

Integration by Parts

Integration by parts is the name given to the product rule for differentiation when it is used as an integration technique. From the differentiation formula

$$\frac{d}{dx}(f(x)g(x)) = f(x)g'(x) + g(x)f'(x),$$

we get the integration formula

$$f(x)g(x) = \int f(x)g'(x)\,dx + \int g(x)f'(x)\,dx,$$

or, equivalently,

$$\int f(x)g'(x)\,dx = f(x)g(x) - \int g(x)f'(x)\,dx. \tag{17.1}$$

Formula (17.1) allows you to trade one integral, $\int f(x)g'(x)\,dx$, for another, $\int g(x)f'(x)\,dx$. The trick is to recognize the cases in which the second integral is simpler than the first.

The utilitarian form of (17.1) is

$$\int u dv = uv - \int v\,du, \tag{17.2}$$

where we have let $u = f(x)$, $dv = g'(x)\,dx$, so $v = g(x)$, and $du = f'(x)\,dx$.

A noteworthy particular case of equation (17.1) is the case $g(x) = 1$, or, in the notation of (2), $dv = dx$. In this case,

$$\int f(x)\,dx = xf(x) - \int xf'(x)\,dx. \tag{17.3}$$

Sometimes $xf'(x)$ is easier to integrate than $f(x)$, and we will use (17.3) to integrate $\log x$ and the inverse trigonometric functions.

EXAMPLE 17.1

$\int \log x \, dx.$

We let $u = \log x$, $dv = dx$, so $v = x$, and $du = \frac{1}{x}\, dx$. Then

$$\begin{aligned} \int \log x \, dx &= x \log x - \int x \cdot \frac{1}{x} \, dx \\ &= x \log x - \int dx \\ &= x \log x - x. \end{aligned} \tag{17.4}$$

Since $\log x$ is one of our standard elementary functions, (17.4) should be considered one of our standard integration formulas. It wouldn't hurt to memorize it.

EXAMPLE 17.2

$\int \sin^{-1} x \, dx.$

Let $u = \sin^{-1} x$, $dv = dx$, so $v = x$ and $du = \frac{1}{\sqrt{1-x^2}}\, dx$. Thus,

$$\begin{aligned} \int \sin^{-1} x \, dx &= x \sin^{-1} x - \int \frac{x}{\sqrt{1-x^2}} \, dx \\ &= x \sin^{-1} x + \sqrt{1-x^2}. \end{aligned}$$

The technique of Example 17.2 also works easily for $\cos^{-1} x$, $\tan^{-1} x$, but these integrals occur infrequently, and it is easier to go through the integration technique or look in the integration table than it is to memorize the formula.

The more usual situation, where dv is not simply dx, is illustrated in the next example.

EXAMPLE 17.3

$\int xe^x \, dx.$

Let $u = x$, $dv = e^x \, dx$, so $v = e^x$ and $du = dx$. Thus,

$$\begin{aligned} \int xe^x \, dx &= xe^x - \int e^x \, dx \\ &= xe^x - e^x. \end{aligned}$$

For definite integrals the integration by parts formula is

$$\int_a^b u \, dv = uv \Big]_a^b - \int_a^b v \, du.$$

EXAMPLE 17.4

$\int_0^2 xe^{x/2} \, dx.$

Here we let $u = x$, $dv = e^{x/2}$, so $v = \frac{1}{2}e^{x/2}$ and $du = dx$. Then we have

$$\begin{aligned} \int_0^2 xe^{x/2} \, dx &= x\left(\frac{1}{2}e^{x/2}\right)\Big]_0^2 - \int_0^2 \frac{1}{2}e^{x/2} \, dx \\ &= 2 \cdot \frac{1}{2} \cdot e - e^{x/2}\Big]_0^2 \\ &= e - (e-1) = 1. \end{aligned}$$

EXAMPLE 17.5
Find the volume swept out by rotating about the y-axis the area bounded by $y = e^{x/2}$, the x-axis, and the line $x = 2$ (Figure 17.1).

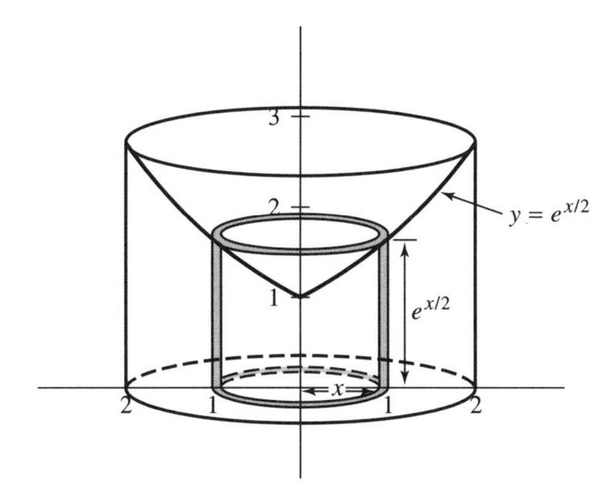

Figure 17.1

This method of calculating volume is called the **cylindrical shells method**. We think of the thin-walled cylindrical shell swept out by the vertical strip $e^{x/2}$ high and dx wide. As this strip moves around a circle of circumference $2\pi x$, it sweeps out a thin cylindrical shell of circumference $2\pi x$, height $e^{x/2}$, and thickness dx. If you cut the shell and flatten it out you get an approximate rectangular solid $2\pi x$ long, $e^{\frac{x}{2}}$ wide, and dx thick. Thus, this increment of volume is

$$dV = 2\pi x \cdot e^{x/2}\, dx.$$

The volume swept out by strips between $x = 0$ and $x = 2$ is therefore, using the result of Example 17.4,

$$\begin{aligned} V &= \int_0^2 2\pi x e^{x/2}\, dx \\ &= 2\pi \int_0^2 x e^{x/2}\, dx \\ &= 2\pi(1). \end{aligned}$$

The following problems, except Problem 17.25, are examples of problems where the integration by parts technique works. Moreover, these problems contain most of the standard examples where integration by parts works. We mention this because many students fall in love with integration by parts and try to use this technique for *every* problem. If your integration problem does not look like one of those in this section, try something else.

PROBLEMS

17.1 $\displaystyle\int \cos^{-1} x\, dx$

17.2 $\displaystyle\int \tan^{-1} x\, dx$

17.3 $\displaystyle\int x e^{-2x}\, dx$

17.4 $\displaystyle\int x^2 e^x\, dx$ (Integrate by parts twice.)

17.5 $\displaystyle\int x^3 e^{-2x}\, dx$

17.6 $\int x \sin x \, dx$

17.7 $\int x \cos 3x \, dx$

17.8 $\int x^2 \sin 5x \, dx$

17.9 $\int x \log x \, dx$

17.10 $\int \sqrt{x} \log x \, dx$

17.11 $\int x^7 \log x \, dx$

17.12 $\int (\log x)^2 \, dx$

17.13 $\int (\log x)^3 \, dx$

17.14 $\int x\sqrt{1+x} \, dx$

17.15 $\int \log(1+x^2) \, dx$

17.16 $\int e^x \sin x \, dx$. (Integrate by parts twice, with $u = e^x$ each time, and solve the resulting equation for the integral.)

17.17 $\int \frac{\log(x+1)}{\sqrt{x+1}} \, dx$

17.18 $\int x^3 e^{-x^2} \, dx$

17.19 $\int_0^{\frac{\pi}{2}} x \cos x \, dx$

17.20 $\int_0^1 x \log(1+x) \, dx$ *Hint:* $\int \frac{x^2 \, dx}{x+1}$ is an easy Chapter 16 problem.

17.21 $\int_0^2 x\sqrt{x+2} \, dx$

17.22 $\int_0^{\frac{1}{4}} \sin^{-1} 2x \, dx$

17.23 Use the cylindrical shells method to find the volume obtained by rotating about the y-axis the area in the first quadrant under $y = \log x$, for $1 \le x \le e$.

17.24 Find the volume obtained by rotating about the y-axis the area in the first quadrant under $y = \sin x$, for $0 \le x \le \frac{\pi}{2}$.

17.25 Use the cylindrical shells method to find the volume of the hemisphere obtained by rotating about the y-axis the area under $y = \sqrt{a^2 - x^2}$ for $0 \le x \le a$. (This is *not* an integration by parts problem.)

17.26 (i) Integrate $\int \sec^3 x \, dx$ by parts using $u = \sec x$, $dv = \sec^2 x \, dx$.
(ii) Use the same idea to integrate $\int \sec^n x \, dx$ to get the reduction formula $\int \sec^n x \, dx = \frac{1}{n-1} \sec^{n-2} x \tan x + (\frac{n-2}{n-1}) \int \sec^{n-2} x \, dx$.

18

Trigonometric Integrals

Integrals involving trigonometric functions arise very frequently, in part because of substitutions that turn radical expressions into trigonometric functions. It is therefore necessary to gain some proficiency in such integrals, and that is our goal in this chapter. We start by reviewing some familiar integrals:

$$\int \sin x \, dx = -\cos x;$$

$$\int \cos x \, dx = \sin x;$$

$$\int \tan x \, dx = \int \frac{\sin x}{\cos x} \, dx = -\log|\cos x| = \log|\sec x|;$$

$$\int \sec^2 x \, dx = \tan x.$$

The following standard trigonometric identities will be used constantly, and should be memorized:

$$\sin^2 x + \cos^2 x = 1; \tag{18.1}$$

$$\tan^2 x + 1 = \sec^2 x; \tag{18.2}$$

$$\begin{cases} \sin(x+y) = \sin x \cos y + \cos x \sin y; \\ \sin(x-y) = \sin x \cos y - \cos x \sin y; \end{cases} \tag{18.3}$$

$$\begin{cases} \cos(x+y) = \cos x \cos y - \sin x \sin y; \\ \cos(x-y) = \cos x \cos y + \sin x \sin y; \end{cases} \tag{18.4}$$

$$\sin^2 x = (1 - \cos 2x)/2; \tag{18.5}$$

$$\cos^2 x = (1 + \cos 2x)/2. \tag{18.6}$$

We deal primarily with $\sin x$, $\cos x$, $\tan x$, and $\sec x$. We have integration formulas for $\sin x$, $\cos x$, and $\tan x$, and the integral of $\sec x$ is the following:

$$\int \sec x \, dx = \log | \sec x + \tan x | \,. \tag{18.7}$$

Formula (18.7) can easily be checked by differentiating the right side:

$$\begin{aligned} \frac{d}{dx} \log | \sec x + \tan x | &= \frac{\sec x \tan x + \sec^2 x}{\sec x + \tan x} \\ &= \frac{\sec x (\tan x + \sec x)}{(\sec x + \tan x)} \\ &= \sec x \,. \end{aligned}$$

If you put integral signs in front of the above formulas and read from down to up, you get the usual "derivation" of formula (18.7).

The simple u-substitution, $u = \sin x$, $du = \cos x \, dx$, works for any integral of the form

$$\int \sin^k x \cos x \, dx = \int u^k du = \frac{1}{k+1} \sin^{k+1} x. \tag{18.8}$$

Here k can be any number except -1; that is, k can be positive or negative, fraction or integer. The u-substitution of (18.8) is so simple that the intermediate formula $\int u^k du$ need not be written down. The following integrals are examples of (18.8):

$$\int \sin^2 x \cos x \, dx = \frac{1}{3} \sin^3 x;$$

$$\int \sqrt{\sin x} \cos x \, dx = \frac{2}{3} \sin^{\frac{3}{2}} x;$$

$$\int \frac{\cos x}{\sin^5 x} dx = -\frac{1}{4} \sin^{-4} x.$$

The same trick works, of course, if the sines and cosines are interchanged; in this case $u = \cos x$, $du = - \sin x \, dx$, and we have

$$\int \cos^k x \sin x \, dx = \int u^k (-1) du = -\frac{1}{k+1} \cos^{k+1} x. \tag{18.9}$$

The following are examples of (18.9):

$$\int \cos^{\frac{3}{2}} x \sin x \, dx = -\frac{2}{5} \cos^{\frac{5}{2}} x;$$

$$\int \cos^{-2} x \sin x \, dx = \frac{1}{\cos x} = \sec x;$$

$$\int \cos^4 x \sin x \, dx = -\frac{1}{5} \cos^5 x.$$

The following example illustrates a simple variation of the $\int \sin^k x \cos x \, dy$ integrals. The trick depends on $\cos x$ occurring to an odd power.

$$\begin{aligned} \int \sin^2 x \cos^3 x \, dx &= \int \sin^2 x \cos^2 x \cos x \, dx \\ &= \int \sin^2 x (1 - \sin^2 x) \cos x \, dx \\ &= \int \sin^2 x \cos x \, dx - \int \sin^4 x \cos x \, dx \\ &= \frac{1}{3} \sin^3 x - \frac{1}{5} \sin^5 x. \end{aligned}$$

This technique works for all integrals of the form $\int \sin^m x \cos^n x\,dx$ where either m or n is a positive odd integer. For example, $(m = 5, n = 0)$,

$$\begin{aligned}\int \sin^5 x\,dx &= \int \sin^4 x \sin x\,dx \\ &= \int (1-\cos^2 x)^2 \sin x\,dx \\ &= \int (1 - 2\cos^2 x + \cos^4 x)\sin x\,dx \\ &= -\cos x + \frac{2}{3}\cos^3 x - \frac{1}{5}\cos^5 x.\end{aligned}$$

Integrals of the form $\int \sin^m x \cos^n x\,dx$ where both m and n are even are handled using formulas (18.5) and (18.6), which express the squares in terms of $\cos 2x$. For example, $(m = 0,\ n = 2)$,

$$\begin{aligned}\int \cos^2 x\,dx &= \int \frac{1+\cos 2x}{2}\,dx \\ &= \frac{1}{2}x + \frac{1}{4}\sin 2x.\end{aligned}$$

It is sometimes more convenient to have the answer expressed in terms of $\sin x$ and $\cos x$ rather than $\sin 2x$. In this case, use (18.3), which gives $\sin 2x = 2\sin x \cos x$, so

$$\frac{1}{2}x + \frac{1}{4}\sin 2x = \frac{1}{2}x + \frac{1}{2}\sin x \cos x.$$

Here is another example, involving both the sine and cosine to an even power:

$$\begin{aligned}\int \sin^4 x \cos^2 x dx &= \int \left(\frac{1-\cos 2x}{2}\right)^2 \left(\frac{1+\cos 2x}{2}\right) dx \\ &= \frac{1}{8}\int (1 - 2\cos 2x + \cos^2 2x)(1+\cos 2x)\,dx \\ &= \frac{1}{8}\int (1 - \cos 2x - \cos^2 2x + \cos^3 2x)\,dx.\end{aligned}$$

The integral $\int \cos^2 x\,dx$ was done above, and $\int \cos^3 x\,dx$ uses the trick for m or n odd:

$$\begin{aligned}\int \cos^3 2x\,dx &= \int \cos^2 2x \cos 2x\,dx \\ &= \int (1 - \sin^2 2x)\cos 2x\,dx \\ &= \frac{1}{2}\sin 2x - \frac{1}{6}\sin^3 2x.\end{aligned}$$

Integrals involving secants and tangents use the substitutions

$$u = \tan x, \quad du = \sec^2 x\,dx, \tag{18.10}$$

or

$$u = \sec x, \quad du = \sec x \tan x\,dx. \tag{18.11}$$

For example, using (18.10), we get

$$\int \tan^{\frac{2}{3}} x \sec^2 x\, dx = \int u^{\frac{2}{3}} du$$
$$= \frac{3}{5} u^{\frac{5}{3}}$$
$$= \frac{3}{5} \tan^{\frac{5}{3}} x.$$

Using (18.11) we have

$$\int \sec^3 x \tan x\, dx = \int \sec^2 x \sec x \tan x\, dx$$
$$= \int u^2 du$$
$$= \frac{1}{3} u^3$$
$$= \frac{1}{3} \sec^3 x.$$

The following integral pops up curiously often, and it will pay you to remember that it is written down here:

$$\int \sec^3 x\, dx = \frac{1}{2}[\sec x \tan x + \log|\sec x + \tan x|]. \tag{18.12}$$

Notice that the integral is the average of the derivative of $\sec x$ and the integral of $\sec x$; this is a coincidence, but it makes the formula easy to remember. It is easy to check (18.12) by differentiating (Problem 18.25).

PROBLEMS

18.1 $\int \sin^3 x \cos x\, dx$

18.2 $\int \sin^{\frac{3}{2}} x \cos x\, dx$

18.3 $\int \cos^5 x \sin x\, dx$

18.4 $\int \frac{\cos x}{\sin x} dx$

18.5 $\int \cos^3 x\, dx = \int (1 - \sin^2 x) \cos x\, dx$

18.6 $\int \sin x \cos^3 x\, dx$

18.7 $\int \sin^3 x\, dx$

18.8 $\int \sqrt{\cos x} \sin x\, dx$

18.9 $\int \cos^2 x \sin^3 x\, dx$

18.10 $\int \sin^2 x dx$. Write the answer in terms of $\sin x$ and $\cos x$.

18.11 $\int \sin^2 x \cos^2 x\, dx$. Write the answer in terms of $\sin x$ and $\cos x$.

18.12 $\int \cos^4 x \, dx$

18.13 $\int \cos^4 x \sin^2 x \, dx$

18.14 $\int \sec^2 2x \, dx$

18.15 $\int \tan^2 x \sec^2 x \, dx$

18.16 $\int \tan^2 x \, dx$

18.17 $\int \sec^4 3x \tan 3x \, dx$

18.18 $\int \frac{\tan x}{\sec x} \, dx$

18.19 $\int \sqrt{\tan x} \sec^2 x \, dx$

18.20 $\int \tan^3 x \, dx = \int (\sec^2 x - 1) \tan x \, dx$

18.21 $\int \frac{1 + \sin x}{\cos x} dx$

18.22 $\int \sec^3 4x \, dx$

18.23 $\int \frac{\sec 2x}{\cos^2 2x} dx$

18.24 $\int \frac{\tan x}{\cos x} dx$

18.25 Verify formula (17.12) by showing that the derivative of $\frac{1}{2}[\sec x \tan x + \log | \sec x + \tan x |]$ is $\sec^3 x$.

18.26 (i) Integrate by parts to derive the reduction formula $\int \sin^n x \, dx = -\frac{1}{n} \sin^{n-1} x \cos x + (\frac{n-1}{n}) \int \sin^{n-2} x \, dx$.
(ii) Use the formula twice to integrate $\int \sin^4 x \, dx$.

18.27 Find the area under $y = \sin^2 x \cos^2 x$ for $0 \le x \le \frac{\pi}{2}$.

18.28 Find the volume obtained by rotating around the x-axis the area under $y = \sin^2 x, \quad 0 \le x \le \pi$.

18.29 Let x_1 be the largest negative number where the curves $y = \cos x$, $y = \sin x$ intersect, and let x_2 be the smallest positive number where the curves intersect. Find the area between $y = \cos x$ and $y = \sin x$ for $x_1 \le x \le x_2$.

18.30 Show that for integers m and n, $\int_0^{2\pi} \sin mx \cos nx \, dx = 0$. *Hint:* $\sin mx \cos nx = \frac{1}{2}[\sin(m+n)x + \sin(m-n)x]$.

18.31 Show that for all integers n, $\int_0^{2\pi} \sin^2 nx \, dx = \int_0^{2\pi} \cos^2 nx \, dx = \pi$.

18.32 Find the indefinite integrals (cf. Problem 18.30).
(i) $\int \sin 3x \cos 2x \, dx$
(ii) $\int \sin x \, \cos 5x \, dx$

19

Trigonometric Substitution

Integrands that contain one of the radical expressions $\sqrt{a^2 - x^2}$, $\sqrt{a^2 + x^2}$, $\sqrt{x^2 - a^2}$ can frequently be simplified by making a trigonometric substitution.

$$\text{For } \sqrt{a^2 - x^2}, \quad \text{let } x = a \sin\theta, \quad dx = a\cos\theta\, d\theta. \tag{19.1}$$

$$\text{For } \sqrt{a^2 + x^2}, \quad \text{let } x = a \tan\theta, \quad dx = a\sec^2\theta\, d\theta. \tag{19.2}$$

$$\text{For } \sqrt{x^2 - a^2}, \quad \text{let } x = a \sec\theta, \quad dx = a\sec\theta\tan\theta\, d\theta. \tag{19.3}$$

With these substitutions the radicals simplify as follows:

$$\sqrt{a^2 - x^2} = \sqrt{a^2 - a^2\sin^2\theta} = a\cos\theta; \tag{19.4}$$

$$\sqrt{a^2 + x^2} = \sqrt{a^2 + a^2\tan^2\theta} = a\sec\theta; \tag{19.5}$$

$$\sqrt{x^2 - a^2} = \sqrt{a^2\sec^2\theta - a^2} = a\tan\theta. \tag{19.6}$$

In these substitutions we assume that $a > 0$ and θ is such that the right sides of (19.4), (19.5), (19.6) are positive.

EXAMPLE 19.1

Find the area of the top half of the circle $x^2 + y^2 = 4$.

The equation of the top half of the circle is $y = \sqrt{4 - x^2}$, and the area is

$$\int_{-2}^{2} \sqrt{4 - x^2}\, dx. \tag{19.7}$$

We make the substitution $x = 2\sin\theta$, $dx = 2\cos\theta\, d\theta$. When $x = -2$, $\sin\theta = -1$ and $\theta = -\frac{\pi}{2}$.

Similarly, when $x = 2$, $\sin\theta = 1$ and $\theta = \frac{\pi}{2}$. Therefore, the new integral is

$$\begin{aligned}
\int_{-2}^{2} \sqrt{4 - x^2}\, dx &= \int_{-\frac{\pi}{2}}^{\frac{\pi}{2}} \sqrt{4 - 4\sin^2\theta}\; 2\cos\theta\, d\theta \\
&= \int_{-\frac{\pi}{2}}^{\frac{\pi}{2}} 2\sqrt{1 - \sin^2\theta}\; 2\cos\theta\, d\theta \\
&= 4\int_{-\frac{\pi}{2}}^{\frac{\pi}{2}} \cos^2\theta\, d\theta \\
&= \frac{4}{2}\int_{-\frac{\pi}{2}}^{\frac{\pi}{2}} (1 + \cos 2\theta)\, d\theta \\
&= 2\left[\theta + \frac{1}{2}\sin 2\theta\right]_{-\frac{\pi}{2}}^{\frac{\pi}{2}} \\
&= 2\left[\frac{\pi}{2} - \left(-\frac{\pi}{2}\right)\right] = 2\pi.
\end{aligned} \tag{19.8}$$

The term $\sin 2\theta$ is zero at both $\frac{\pi}{2}$ and $-\frac{\pi}{2}$. As expected, the answer is $\frac{1}{2}\pi(2)^2 = 2\pi$.

EXAMPLE 19.2

$$\int \sqrt{4 - x^2}\, dx.$$

The integrand here is the same as in Example 19.1, but here we want an antiderivative —*a* function of x— rather than a number. Make the same substitution as in Example 19.1, ignoring the limits, and follow the calculations down to the fifth line:

$$\int \sqrt{4 - x^2}\, dx = 2\left[\theta + \frac{1}{2}\sin 2\theta\right].$$

To turn the right side back into a function of x, we use $\theta = \sin^{-1}\frac{x}{2}$, and $\sin 2\theta = 2\sin\theta\cos\theta$; thus,

$$\int \sqrt{4 - x^2}\, dx = 2[\theta + \sin\theta\cos\theta].$$

From Figure 19.1 we see that if $\sin\theta = \frac{x}{2}$, then $\cos\theta = \sqrt{4 - x^2}/2$, and so

$$\begin{aligned}
\int \sqrt{4 - x^2}\, dx &= 2\left[\sin^{-1}\frac{x}{2} + \frac{x}{2}\cdot\frac{\sqrt{4 - x^2}}{2}\right] \\
&= 2\sin^{-1}\frac{x}{2} + \frac{1}{2}x\sqrt{4 - x^2}.
\end{aligned}$$

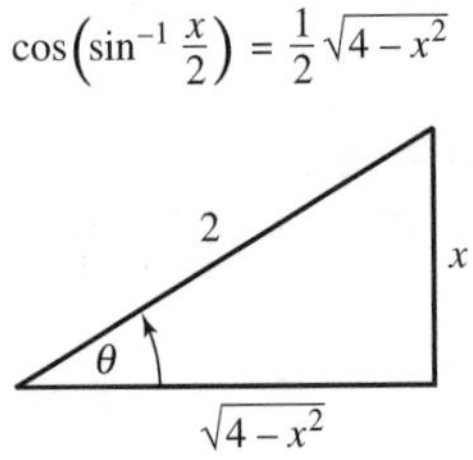

$\cos\left(\sin^{-1}\frac{x}{2}\right) = \frac{1}{2}\sqrt{4 - x^2}$

2

x

θ

$\sqrt{4 - x^2}$

Figure 19.1

EXAMPLE 19.3

$$\int_{0}^{3} \sqrt{9 + x^2}\, dx.$$

Here we let $x = 3\tan\theta$, $dx = 3\sec^2\theta\, d\theta$. When $x = 0$, $\tan\theta = 0$ so $\theta = 0$. When $x = 3$, $\tan\theta = 1$ so

$\theta = \frac{\pi}{4}$. Therefore, we have

$$\begin{aligned}\int_0^3 \sqrt{9+x^2}\,dx &= \int_0^{\frac{\pi}{4}} \sqrt{9+9\tan^2\theta}\;3\sec^2\theta\,d\theta \\ &= \int_0^{\frac{\pi}{4}} 9\sqrt{\sec^2\theta}\;\sec^2\theta\,d\theta \\ &= 9\int_0^{\frac{\pi}{4}} \sec^3\theta\,d\theta.\end{aligned}$$

Using equation (18.12) from the last chapter,

$$\begin{aligned}9\int_0^{\frac{\pi}{4}} \sec^3\theta\,d\theta &= \frac{9}{2}\left[\sec\theta\tan\theta + \log|\sec\theta+\tan\theta|\right]_0^{\frac{\pi}{4}} \\ &= \frac{9}{2}\left[\sec\frac{\pi}{4}\tan\frac{\pi}{4} + \log\left|\sec\frac{\pi}{4}+\tan\frac{\pi}{4}\right| - \sec 0\tan 0 - \log|\sec 0+\tan 0|\right] \\ &= \frac{9}{2}\left[\sqrt{2}\cdot 1 + \log\left|\sqrt{2}+1\right| - 0 - 0\right] \\ &= \frac{9}{2}\left(\sqrt{2} + \log\left|\sqrt{2}+1\right|\right).\end{aligned}$$

EXAMPLE 19.4

$$\int \sqrt{9+x^2}\,dx.$$

This is the same integrand as Example 19.3, but now we want an antiderivative. With the same substitution, $x = 3\tan\theta$, and the same calculations we arrive at

$$\begin{aligned}\int \sqrt{9+x^2}\,dx &= 9\int \sec^3\theta\,d\theta \\ &= \frac{9}{2}[\sec\theta\tan\theta + \log|\sec\theta+\tan\theta|].\end{aligned}$$

From Figure 19.2 we see that if $\tan\theta = \frac{x}{3}$, then $\sec\theta = \sqrt{9+x^2}/3$. Hence,

$$\begin{aligned}\int \sqrt{9+x^2}\,dx &= \frac{9}{2}\left[\frac{\sqrt{9+x^2}}{3}\cdot\frac{x}{3} + \log\left|\frac{\sqrt{9+x^2}}{3}+\frac{x}{3}\right|\right] \\ &= \frac{1}{2}x\sqrt{9+x^2} + \frac{9}{2}\log\left|\sqrt{9+x^2}+x\right| - \frac{9}{2}\log 3.\end{aligned}$$

The last step uses

$$\log\left|\frac{\sqrt{9+x^2}+x}{3}\right| = \log\left|\sqrt{9+x^2}+x\right| - \log 3.$$

We can discard the constant $-\frac{9}{2}\log 3$ since antiderivatives are unique only up to an additive constant, and we write the antiderivative as

$$\int \sqrt{9+x^2}\,dx = \frac{1}{2}x\sqrt{9+x^2} + \frac{9}{2}\log\left|\sqrt{9+x^2}+x\right|.$$

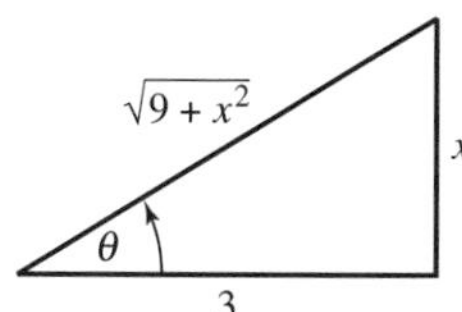

Figure 19.2

EXAMPLE 19.5

$$\int \frac{\sqrt{x^2-1}}{x}\,dx.$$

Here the substitution is $x = \sec\theta$, $dx = \sec\theta\tan\theta\,d\theta$. Since we want an antiderivative as a function of x, we will need the triangle of Figure 19.3. With the substitution we get

$$\begin{aligned}\int \frac{\sqrt{x^2-1}}{x}\,dx &= \int \frac{\sqrt{\sec^2\theta-1}}{\sec\theta}\sec\theta\tan\theta\,d\theta \\ &= \int \tan^2\theta\,d\theta \\ &= \int (\sec^2\theta-1)\,d\theta \\ &= \tan\theta-\theta \\ &= \sqrt{x^2-1}-\sec^{-1}x.\end{aligned}$$

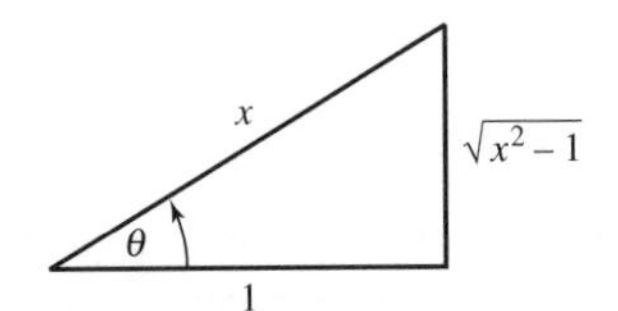

Figure 19.3

EXAMPLE 19.6

$$\int \frac{dx}{(a^2+x^2)^2}.$$

This integral does not involve a radical, but it is an important example of a rational function of a type we did not cover previously. We let $x = a\tan\theta$, $dx = a\sec^2\theta\,d\theta$.

$$\begin{aligned}\int \frac{dx}{(a^2+x^2)^2} &= \int \frac{a\sec^2\theta\,d\theta}{[a^2(1+\tan^2\theta)]^2} \\ &= \frac{1}{a^3}\int \frac{\sec^2\theta\,d\theta}{\sec^4\theta} \\ &= \frac{1}{a^3}\int \cos^2\theta\,d\theta \\ &= \frac{1}{a^3}\frac{1}{2}\left[\theta+\frac{1}{2}\sin 2\theta\right] \\ &= \frac{1}{2a^3}[\theta+\sin\theta\cos\theta].\end{aligned}$$

From Figure 19.4 we see that if $\tan\theta = x/a$, then $\sin\theta = x/\sqrt{a^2+x^2}$ and $\cos\theta = a/\sqrt{x^2+a^2}$. Hence,

$$\int \frac{dx}{(a^2+x^2)^2} = \frac{1}{2a^3}\left[\tan^{-1}\frac{x}{a}+\frac{ax}{x^2+a^2}\right]. \tag{19.9}$$

$$\sec\left(\tan^{-1}\frac{x}{a}\right) = \frac{x}{\sqrt{a^2+x^2}}; \quad \cos\left(\tan^{-1}\frac{x}{a}\right) = \frac{a}{\sqrt{a^2+x^2}}$$

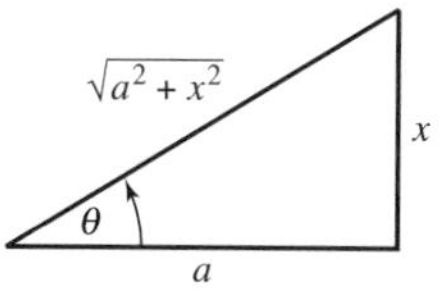

$\sqrt{a^2+x^2}$ x θ a

Figure 19.4

Recall the definitions of the hyperbolic cosine and hyperbolic sine:

$$\cosh x = \frac{1}{2}(e^x + e^{-x}); \quad \sinh x = \frac{1}{2}(e^x - e^{-x}).$$

The hyperbolic functions can also be used to simplify integrals involving radicals, using the following identities, which are similar to the familiar trigonometric identities.

$$\frac{d}{dx}\cosh x = \sinh x; \quad \frac{d}{dx}\sinh x = \cosh x; \tag{19.10}$$

$$\cosh^2 x - \sinh^2 x = 1; \tag{19.11}$$

$$\cosh^2 x = \frac{1}{2}(1 + \cosh 2x); \tag{19.12}$$

$$\sinh 2x = 2\sinh x \cosh x. \tag{19.13}$$

The function $\sinh x$ is strictly increasing on the whole line, and its inverse is denoted $\sinh^{-1} x$. The function $\cosh x$ is increasing on $[0, \infty)$, and we let $\cosh^{-1} x$ denote the non-negative number y such that $\cosh y = x$. The functions $\sinh^{-1} x$, $\cosh^{-1} x$ can be expressed in terms of the logarithm, as we show next. Let $y = \sinh^{-1} x$, so $\sinh y = x$; then we have

$$\frac{1}{2}(e^y - e^{-y}) = x,$$

$$e^y - 2x - e^{-y} = 0,$$

$$e^{2y} - 2xe^y - 1 = 0.$$

The last equation is quadratic in e^y, so

$$e^y = \frac{2x \pm \sqrt{4x^2+4}}{2} = x \pm \sqrt{x^2+1}. \tag{19.14}$$

Since $e^y > 0$ and $x - \sqrt{x^2+1}$ is negative, we must have the plus sign in (19.14), and

$$\begin{aligned} e^y &= x + \sqrt{x^2+1}, \\ y = \sinh^{-1} x &= \log(x + \sqrt{x^2+1}). \end{aligned} \tag{19.15}$$

EXAMPLE 19.7

$\int \sqrt{a^2+x^2}dx.$

We let $x = a \sinh u$, $dx = a \cosh u \, du$, and use the identities (19.10), (19.11), (19.12), (19.13).

$$\begin{aligned}
\int \sqrt{a^2+x^2}\,dx &= \int \sqrt{a^2(1+\sinh^2 u)}\, a\cosh u \, du \\
&= \int a^2 \cosh^2 u \, du \\
&= \frac{a^2}{2}\int (1+\cosh 2u)\, du \\
&= \frac{a^2}{2}\left[u + \frac{1}{2}\sinh 2u\right] \\
&= \frac{a^2}{2}[u + \sinh u \cosh u] \\
&= \frac{a^2}{2}\left[\sinh^{-1}\left(\frac{x}{a}\right) + \frac{x}{a}\sqrt{1+\left(\frac{x}{a}\right)^2}\right] \\
&= \frac{a^2}{2}\left[\log\left(\frac{x}{a} + \sqrt{\left(\frac{x}{a}\right)^2 + 1}\right) + \frac{x}{a}\frac{\sqrt{a^2+x^2}}{a}\right] \\
&= \frac{a^2}{2}\left[\log\left(\frac{x+\sqrt{x^2+a^2}}{a}\right) + \frac{x\sqrt{a^2+x^2}}{a^2}\right] \\
&= \frac{a^2}{2}\log(x+\sqrt{x^2+a^2}) + \frac{1}{2}x\sqrt{a^2+x^2}.
\end{aligned}$$

In the last step we discarded the constant $(\frac{a^2}{2})(-\log a)$.

PROBLEMS

19.1 $\displaystyle\int_{-1}^{1} \sqrt{1-x^2}\,dx$

19.2 $\displaystyle\int \sqrt{1-x^2}\,dx$

19.3 $\displaystyle\int \frac{x^2dx}{\sqrt{4-x^2}}$

19.4 $\displaystyle\int_0^1 x\sqrt{1-x^2}\,dx$

19.5 $\displaystyle\int \frac{xdx}{\sqrt{1-x^2}}$ Do this two ways—trigonometric substitution and u-substitution.

19.6 $\displaystyle\int \frac{x^2dx}{(1-x^2)^{\frac{3}{2}}}$

19.7 $\displaystyle\int \frac{dx}{(9-x^2)^{\frac{3}{2}}}$

19.8 $\displaystyle\int \frac{dx}{\sqrt{3-x^2}}$ Let $x = \sqrt{3}\sin\theta$.

19.9 $\displaystyle\int \frac{dx}{x\sqrt{a^2-x^2}}$ Use $x = a\cos\theta$.

19.10 Find the area inside the ellipse $\frac{x^2}{a^2} + \frac{y^2}{b^2} = 1$ (i.e., twice the area under the curve $y = \frac{b}{a}\sqrt{a^2-x^2}$).

19.11 $\displaystyle\int \frac{dx}{\sqrt{1+x^2}}$

19.12 $\displaystyle\int \frac{x\,dx}{\sqrt{4+x^2}}$ Do this two ways.

19.13 $\displaystyle\int \frac{dx}{x^2\sqrt{1+x^2}}$

19.14 $\displaystyle\int \frac{\sqrt{x^2-4}}{x}\,dx$

19.15 $\displaystyle\int \frac{dx}{x\sqrt{x^2-4}}$

19.16 $\displaystyle\int \frac{x^2dx}{\sqrt{x^2-9}}$

19.17 $\displaystyle\int \sqrt{x^2-1}\,dx$

19.18 $\displaystyle\int \frac{dx}{(1+x^2)^2}$

19.19 $\displaystyle\int \frac{x+1}{(1+x^2)^2}\,dx$ Write as the sum of two fractions, and integrate the two fractions separately.

19.20 The answer in Example 19.15 involves $\sec^{-1} x$, which we haven't used before. Use Figure 19.3 to show that $\sec^{-1} x = \tan^{-1}\sqrt{x^2-1}$. Differentiate $\sec^{-1} x$ as usual by letting $y = \sec^{-1} x$, $\sec y = x$. Then compare with $\frac{d}{dx}\tan^{-1}\sqrt{x^2-1}$.

19.21 A small weight at $(a, 0)$ is attached to a string of length a that initially lies along the x-axis from 0 to a. The end of the string is moved up the y-axis, pulling the weight along a curve called a **tractrix**. Find the equation of the curve. *Hint*: When the weight is at (x, y), the string is the hypoteneuse of a right triangle whose horizontal leg has length x and whose vertical leg along the y-axis has length $\sqrt{a^2-x^2}$. Therefore, $\frac{dy}{dx} = -\frac{\sqrt{a^2-x^2}}{x}$. Integrate to get y as a function of x. *Note*: If you prefer more exciting imagery, imagine (x, y) to be the position of a fast attack submarine, tracking its target at a constant distance, a, while the target sails due north along the y-axis.

19.22 Verify (19.11), (19.12), (19.13).

19.23 Show that $\sinh^2 x = \dfrac{1}{2}[\cosh 2x - 1]$.

19.24 Let (x, y) be a point on the curve formed by a flexible cable that is suspended at two points—for example, a telephone line. Assume the low point on the curve is at $(0, a)$. If s is the weight of the line between $(0, a)$ and (x, y), then parametric equations of the curve are

$$x = a\sinh^{-1}\frac{s}{a}; \quad y = \sqrt{a^2+s^2}.$$

Solve both equations for $\frac{s^2}{a^2}$ and eliminate the parameter s. Show the Cartesian equation is $y = a\cosh\frac{x}{a}$. (Such a curve is called a **catenary**.)

19.25 (i) Use a trigonometric substitution to show

$$\int \frac{dx}{\sqrt{a^2+x^2}} = \log(x+\sqrt{x^2+a^2}).$$

(ii) Use a hyperbolic substitution to show

$$\int \frac{dx}{\sqrt{a^2+x^2}} = \sinh^{-1}\frac{x}{a}.$$

(iii) Use (19.15) to show that the two answers differ by a constant.

19.26 Find a formula for $\cosh^{-1} u$ in terms of the logarithm. *Hint*: The ambiguity of the plus or minus sign is solved by noticing that $u + \sqrt{u^2 - 1} = \frac{1}{u - \sqrt{u^2 - 1}}$, so $\log(u + \sqrt{u^2 - 1}) = -\log(u - \sqrt{u^2 - 1})$. For any given $u > 1$, there are two numbers x and $-x$ such that $\cosh x = \cosh(-x) = u$. By definition, $\cosh^{-1} u$ is the positive one of these numbers.

19.27 Show that $x = a \cosh t$, $y = b \sinh t$ are parametric equations of a hyperbola. (Recall that $x = a \cos t$, $y = b \sin t$ are the parametric equations of the ellipse $\frac{x^2}{a^2} + \frac{y^2}{b^2} = 1$.)

20

Numerical Integration

We have devoted several chapters to techniques for finding antiderivatives, and we can, indeed, find antiderivatives of many common functions. However, many simple functions do not have antiderivatives that can be written in terms of our familiar formulas. For example, e^{x^2}, $\frac{\sin x}{x}$, $\sqrt{1-x^3}$ are functions for which no integration technique will work. In spite of this, we might well need to know a numerical answer for definite integrals such as

$$\int_0^1 e^{x^2}\,dx, \quad \int_{\frac{\pi}{4}}^{\frac{\pi}{2}} \left(\frac{\sin x}{x}\right) dx, \quad \int_0^1 \sqrt{1-x^3}\,dx.$$

To get an approximate answer for integrals like these, or indeed for any definite integral, we can go back to the definition of the definite integral as the limit of Riemann sums of the form

$$\sum_{i=1}^{n} f(c_i)(x_i - x_{i-1}). \tag{20.1}$$

Here $\{x_0, x_1, \ldots, x_n\}$ is a partition of the integration interval, and c_i is a point chosen from the ith subinterval.

We will calculate a Riemann sum approximation for the simple integral $\int_1^3 \frac{1}{x}\,dx$. We know the answer is log 3, and this will allow us to see what kind of accuracy we get. Partition the interval $[1, 3]$ into eight equal subintervals with the points $\{1, \frac{5}{4}, \frac{6}{4}, \frac{7}{4}, \frac{8}{4}, \frac{9}{4}, \frac{10}{4}, \frac{11}{4}, \frac{12}{4}\}$. We will use the midpoints for the c_i, so $c_1 = \frac{9}{8}$, $c_2 = \frac{11}{8}$, $c_3 = \frac{13}{8}, \ldots, c_8 = \frac{23}{8}$, and $x_i - x_{i-1} = \frac{1}{4}$ for all i. Now calculate:

$$\begin{aligned}\sum_{i=1} f(c_i)(x_i - x_{i-1}) &= \frac{1}{4}\left[\frac{8}{9} + \frac{8}{11} + \frac{8}{13} + \cdots + \frac{8}{23}\right] \\ &= 2\left[\frac{1}{9} + \frac{1}{11} + \frac{1}{13} + \cdots + \frac{1}{23}\right] \\ &= 2(.54816) = 1.09632.\end{aligned}$$

The eight-place approximation to log 3 is 1.0986123, so our error is about .002.

The so-called *midpoint rule* used here effectively uses a polygonal path joining the points $(x_0, f(x_0)), (x_1, f(x_1)), \ldots, (x_n, f(x_n))$ to approximate the graph of $f(x)$. Since most graphs are curved rather than straight, we will get much better results by approximating the graph of $f(x)$ by segments of parabolas. Parabolas are curved, and they are algebraically the simplest curves next to straight lines.

Here is the idea: We divide up the integration interval $[a, b]$ into an even number, n, of equal subintervals with the partition $\{x_0, x_1, \ldots, x_n\}$. Let $y_i = f(x_i)$, so the curve passes through the points $(x_0, y_0), (x_1, y_1), \ldots, (x_n, y_n)$. There is exactly one parabola through the first three points (x_0, y_0), (x_1, y_1), (x_2, y_2), and we can find the area under that parabola exactly. If the intervals $[x_{i-1}, x_i]$ are small, then the parabola will closely approximate the curve $y = f(x)$, and the area under the parabola will approximate $\int_{x_0}^{x_2} f(x)\,dx$. We do the same thing with the next three points (x_2, y_2), (x_3, y_3), (x_4, y_4), and so on, and we use the total area under the $\frac{n}{2}$ parabolas as our approximation to $\int_a^b f(x)\,dx$. (See Figure 20.1.)

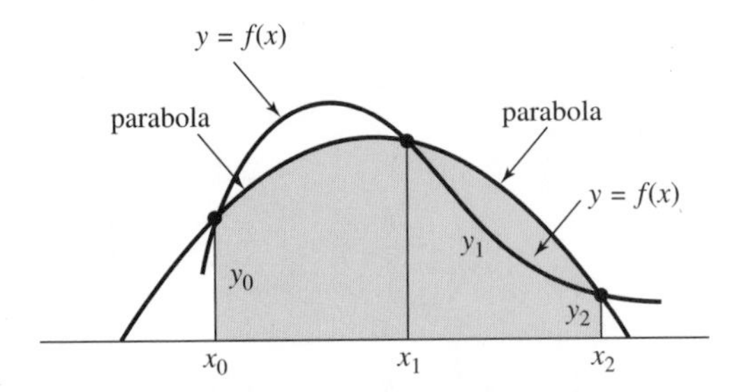

Figure 20.1

Suppose we partition $[a, b]$ into n subintervals, each of length $h = \frac{(b-a)}{n}$. To find the area under the parabola through (x_0, y_0), (x_1, y_1), (x_2, y_2), we move the three points over so that the middle point lies on the y-axis. This simplifies the calculation but doesn't change the area. The three points are then $(-h, y_0)$, $(0, y_1)$, (h, y_2). It is easy to check (Problem 20.6) that the parabola through these points is

$$p(x) = \frac{1}{2}(y_0 + y_2 - 2y_1)\frac{x^2}{h^2} + \frac{1}{2}(y_2 - y_0)\frac{x}{h} + y_1, \tag{20.2}$$

and the area under $p(x)$ is

$$\int_{-h}^{h} p(x)\,dx = \frac{h}{3}[y_0 + 4y_1 + y_2]. \tag{20.3}$$

Now add up the areas under the $\frac{n}{2}$ parabolas (and you see why n must be even) to get Simpson's Rule for n points:

$$\begin{aligned} S_n &= \frac{h}{3}[y_0 + 4y_1 + y_2] + \frac{h}{3}[y_2 + 4y_3 + y_4] + \cdots + \frac{h}{3}[y_{n-2} + 4y_{n-1} + y_n] \\ &= \frac{h}{3}[y_0 + 4y_1 + 2y_2 + 4y_3 + 2y_4 + \cdots + 4y_{n-1} + y_n]. \end{aligned} \tag{20.4}$$

We will use Simpson's Rule with $n = 8$ to approximate $\int_1^3 \frac{1}{x}\,dx$, and compare the result with our earlier estimate. Here $h = \frac{1}{4}$, and the partition points are again $1, \frac{5}{4}, \frac{6}{4}, \ldots, \frac{12}{4}$ so $y_0, \ldots, y_n$ are the reciprocals

$$1, \frac{4}{5}, \frac{4}{6}, \frac{4}{7}, \ldots, \frac{4}{12}.$$

Hence,

$$S_8 = \frac{1}{3} \cdot \frac{1}{4}\left[1 + 4\left(\frac{4}{5}\right) + 2\left(\frac{4}{6}\right) + 4\left(\frac{4}{7}\right) + 2\left(\frac{4}{8}\right) + 4\left(\frac{4}{9}\right) + 2\left(\frac{4}{10}\right) + 4\left(\frac{4}{11}\right) + \frac{4}{12}\right]$$
$$= \frac{1}{3}\left[\frac{1}{4} + \frac{4}{5} + \frac{2}{6} + \frac{4}{7} + \frac{2}{8} + \frac{4}{9} + \frac{2}{10} + \frac{4}{11} + \frac{1}{12}\right]$$
$$= \frac{1}{3}(3.296176) = 1.09873.$$

Comparing this with $\log 3 \doteq 1.0986123$, we see that the error is about .0001, considerably less than the .002 error with the midpoint rule.

Suppose we do not know the "exact" value of $\int_1^3 \frac{1}{x}\,dx$, so we do not know how good an approximation S_8 is. We then try a larger n and smaller h to see how the approximation changes. For example, we compute $S_{10}(h = 0.2)$ and $S_{20}(h = 0.1)$ and compare with S_8:

$$S_8 = 1.0987253,$$
$$S_{10} = 1.0986606,$$
$$S_{20} = 1.0986155.$$

The error in S_n is less than $\frac{K}{n^4}$, where K is a constant that depends on the function $f(x)$ and the interval $[a, b]$. It follows that doubling n will cut the error by a factor of 16 and thus provide at least one more decimal place of accuracy. Looking at S_8 and S_{10} as given above, we would guess that $\log 3 = 1.0986$ or 1.0987. Knowing that S_{20} is one decimal place more accurate than S_{10}, we can then be pretty sure of 1.0986.

In many cases, a reasonable decimal approximation to a definite integral can be obtained by Simpson's Rule with less effort than finding an antiderivative and evaluating it. We illustrate with the following integral, which is Example 19.3 of the last chapter:

$$\int_0^3 \sqrt{9 + x^2}\,dx.$$

With the substitution $x = 3\tan\theta$, we arrive, after much calculation, at

$$\int_0^3 \sqrt{9 + x^2}\,dx = 9\int_0^{\frac{\pi}{4}} \sec^3\theta\,d\theta$$
$$= \frac{9}{2}[\sec\theta\tan\theta + \log|\sec\theta + \tan\theta|]_0^{\frac{\pi}{4}}$$
$$= \frac{9}{2}[\sqrt{2} + \log|1 + \sqrt{2}|].$$

There are many time-consuming steps left out of the above outline. Now to get a decimal answer, we consult the calculator and find

$$\int_0^3 \sqrt{9 + x^2}\,dx \doteq 10.330142.$$

Now use Simpson's Rule with, say, $n = 6$, $h = .5$, and we get

$$S_6 = \frac{.5}{3}\Big[\sqrt{9} + 4\sqrt{9 + .5^2} + 2\sqrt{9 + 1^2} +$$
$$+ 4\sqrt{9 + 1.5^2} + 2\sqrt{9 + 2.0^2} + 4\sqrt{9 + 2.5^2} + \sqrt{9 + 3^2}\Big]$$
$$= 10.330122.$$

This is close enough for most purposes and requires very little effort.

PROBLEMS

20.1 Estimate $\log 2 = \int_1^2 \frac{dx}{x}$ using S_4 and S_{10}.

20.2 Find $S_{10}(h = \frac{1}{10})$ for $\int_0^1 \frac{dx}{1+x^2} = \tan^{-1} 1 = \frac{\pi}{4}$.

20.3 Find S_{10} for $\int_0^1 \sqrt{1-x^2}\, dx$. (This is the area of one-quarter of the unit circle, so the answer is $\frac{\pi}{4}$.)

20.4 (a) Find an exact answer for $\int_0^1 \sqrt{9+x^2}\, dx$, using the substitution $\int_0^1 \sqrt{9+x^2}\, dx = \int_0^{\tan^{-1}\frac{1}{3}} 9\sec^3\theta\, d\theta$, and the 1, 3, $\sqrt{10}$ triangle to express $\sec\theta$ when $\tan\theta = \frac{1}{3}$.
(b) Approximate the exact answer of (a) as a decimal.
(c) Calculate S_{10} for the integral.

20.5 Find $S_4(h = \frac{\pi}{8})$ and $S_8(h = \frac{\pi}{16})$ for $\int_0^{\frac{\pi}{2}} \sin x^2\, dx$. (There is no elementary antiderivative for $\sin x^2$.)

20.6 (a) Verify that the quadratic function $p(x)$ of (20.2) goes through $(-h, y_0)$, $(0, y_1)$, (h, y_2).
(b) Verify that $\int_{-h}^{h} p(x)\, dx = \frac{h}{3}[y_0 + 4y_1 + y_2]$.

20.7 Explain why S_n, for any n, gives the exact answer to $\int_a^b (Ax^2+Bx+C)\, dx$, for any quadratic function and any interval. Start with $n = 2$.

20.8 (i) Check that S_2 (with $h = \frac{(b-a)}{2}$, $x_0 = a$, $x_1 = \frac{(a+b)}{2}$, $x_2 = b$) gives the exact answer to $\int_a^b x^3\, dx$.
(ii) Show that Simpson's Rule with $n = 2$ gives the exact answer for $\int_a^b Q(x)\, dx$ for any cubic polynomial $Q(x)$, and any interval $[a, b]$.

21

Limits at ∞; Sequences

In this chapter we study the limiting behavior of functions $f(x)$ as $x \longrightarrow \infty$. We write $\lim_{x\to\infty} f(x) = L$, or $f(x) \longrightarrow L$ as $x \longrightarrow \infty$, provided the difference $| f(x) - L |$ becomes arbitrarily small for all sufficiently large x. If the function $f(x)$ becomes arbitrarily large, we write $\lim_{x\to\infty} f(x) = \infty$, or $f(x) \longrightarrow \infty$ as $x \longrightarrow \infty$. The usual rules for limits apply, so the limit of a sum is the sum of the limits, and so on.

The limit

$$\lim_{x\to\infty} \frac{x}{x+1} = 1$$

is an obvious example of the kind of behavior we consider. More generally, limits of rational functions at ∞ are all easy to evaluate simply by dividing both numerator and denominator by the highest power of x which occurs in either.

EXAMPLE 21.1

(a) $\lim_{x\to\infty} \frac{x^2-3x}{4x^3-2x+1}$; (b) $\lim_{x\to\infty} \frac{5x^4+3x^2}{7x^4+100x}$; (c) $\lim_{x\to\infty} \frac{x^2+2x}{3x+5}$.

In part (a) we divide top and bottom by x^3:

$$\lim_{x\to\infty} \frac{x^2-3x}{4x^3-2x+1} = \lim_{x\to\infty} \frac{\frac{1}{x} - \frac{3}{x^2}}{4 - \frac{2}{x^2} + \frac{1}{x^3}} = 0.$$

Since the numerator on the right tends to 0 and the denominator to 4, the limit is 0.

In (b) we divide top and bottom by x^4:

$$\lim_{x\to\infty} \frac{5x^4+3x^2}{7x^4+100x} = \lim_{x\to\infty} \frac{5+\frac{3}{x^2}}{7+\frac{100}{x^3}} = \frac{5}{7}.$$

In (c) we divide top and bottom by x^2:

$$\lim_{x\to\infty} \frac{x^2+2x}{3x+5} = \lim_{x\to\infty} \frac{1+\frac{2}{x}}{\frac{3}{x}+\frac{5}{x^2}} = \infty.$$

Since the numerator tends to 1 and the denominator to 0, the "limit" is ∞.

It is clear from this example that if $P(x)$ and $Q(x)$ are polynomials, then $\frac{P(x)}{Q(x)} \longrightarrow 0$ if $Q(x)$ has larger degree than $P(x)$, and $\frac{P(x)}{Q(x)} \longrightarrow \infty$ if $P(x)$ has larger degree than $Q(x)$, and the lead coefficients of $P(x)$ and $Q(x)$ have the same sign. If $P(x)$ and $Q(x)$ have the same degree, then $\frac{P(x)}{Q(x)} \longrightarrow \frac{a_n}{b_n}$ where a_n and b_n are the lead coefficients of $P(x)$ and $Q(x)$.

If a limit isn't obvious, then it is probably one of the following **indeterminate forms**: $\frac{\infty}{\infty}$, $\frac{0}{0}$, $\infty \cdot 0$, 1^∞, or ∞^0. For example,

$$\text{(i)} \quad \lim_{x\to\infty} \frac{\log x}{x} \text{ has the form } \frac{\infty}{\infty};$$

$$\text{(ii)} \quad \lim_{x\to\infty} \frac{\frac{1}{x}}{\frac{1}{\log x}} \text{ has the form } \frac{0}{0};$$

$$\text{(iii)} \quad \lim_{x\to\infty} x \log\left(1 + \frac{1}{x}\right) \text{ has the form } \infty \cdot 0;$$

$$\text{(iv)} \quad \lim_{x\to\infty} \left(1 + \frac{1}{x}\right)^x \text{ has the form } 1^\infty;$$

$$\text{(v)} \quad \lim_{x\to\infty} x^{\frac{1}{x}} \text{ has the form } \infty^0.$$

Notice that (i) and (ii) are exactly the same limit, and this illustrates how $\frac{0}{0}$ forms can be rewritten as $\frac{\infty}{\infty}$ and vice versa. Our standard weapon against the indeterminate forms is l'Hospital's Rule. Since l'Hospital's Rule applies only to the forms $\frac{0}{0}$ and $\frac{\infty}{\infty}$, the others must first be put in one of these forms. For example, the $\infty \cdot 0$ form of (iii) can be put in the $\frac{0}{0}$ form as follows:

$$\lim_{x\to\infty} x \log\left(1 + \frac{1}{x}\right) = \lim_{x\to\infty} \frac{\log\left(1 + \frac{1}{x}\right)}{\frac{1}{x}}.$$

The forms 1^∞ and ∞^0 are treated by considering the limit of the logarithm. Taking the logarithm changes both 1^∞ and ∞^0 to the form $\infty \cdot 0$ (or $0 \cdot \infty$), which can then be treated like (iii) above.

l'Hospital's Rule for $\frac{0}{0}$ or $\frac{\infty}{\infty}$ as $x \longrightarrow \infty$: If both $f(x)$ and $g(x)$ approach zero as $x \longrightarrow \infty$, or if both functions have an infinite limit, then $\lim_{x\to\infty} \frac{f(x)}{g(x)} = \lim_{x\to\infty} \frac{f'(x)}{g'(x)}$ provided that $\frac{f'(x)}{g'(x)}$ approaches a finite limit or tends to infinity, that is, that $\lim_{x\to\infty} \frac{f'(x)}{g'(x)} = L$ or $\lim_{x\to\infty} \frac{f'(x)}{g'(x)} = \infty$.

EXAMPLE 21.2

$\lim_{x\to\infty} \frac{\log x}{x}$.

This has the form $\frac{\infty}{\infty}$ so l'Hospital's Rule will apply if $\frac{f'(x)}{g'(x)}$ has a limit. Here

$$\lim_{x\to\infty} \frac{f'(x)}{g'(x)} = \frac{\frac{1}{x}}{1} = 0.$$

so

$$\lim_{x\to\infty} \frac{\log x}{x} = \lim_{x\to\infty} \frac{\frac{1}{x}}{1} = 0.$$

It is not really necessary to check first that $\lim_{x\to\infty} \frac{f'(x)}{g'(x)}$ makes sense, for that becomes clear in the computation; just write

$$\lim_{x\to\infty} \frac{f(x)}{g(x)} = \lim_{x\to\infty} \frac{f'(x)}{g'(x)}.$$

EXAMPLE 21.3

$\lim_{x\to\infty} x \log\left(1+\frac{1}{x}\right)$.

First put this in the form $\frac{0}{0}$:

$$\lim_{x\to\infty} x \log\left(1+\frac{1}{x}\right) = \lim_{x\to\infty} \frac{\log\left(1+\frac{1}{x}\right)}{\frac{1}{x}}$$

$$= \lim_{x\to\infty} \frac{\frac{1}{1+\frac{1}{x}}\left(-\frac{1}{x^2}\right)}{-\frac{1}{x^2}}$$

$$= \lim_{x\to\infty} \frac{1}{1+\frac{1}{x}} = 1.$$

A **sequence** is a function that is defined only for positive integers, and we will be interested principally in limits at ∞ of sequences. We use the notation $\{x_n\}$ to denote the sequence whose values are $x_1, x_2, x_3, \ldots$. For sequences, x_n is used in place of the usual functional notation $x(n)$. A sequential limit will be indicated by $\lim_{n\to\infty} x_n = L$, or $x_n \longrightarrow L$ as $n \longrightarrow \infty$.

If $f(x) \longrightarrow L$ as $x \longrightarrow \infty$, and we let $x_n = f(n)$, then clearly $x_n \longrightarrow L$ as $n \longrightarrow \infty$. For example, from Example 21.1, we have

$$\lim_{n\to\infty} \frac{n^2 - 3n}{4n^3 - 2n + 1} = 0,$$

and Example 21.2 shows that

$$\lim_{n\to\infty} \frac{\log n}{n} = 0.$$

From Example 21.3 we get

$$\lim_{n\to\infty} n \log\left(1+\frac{1}{n}\right) = \lim_{n\to\infty} \log\left(1+\frac{1}{n}\right)^n = 1,$$

and since the logarithm of $(1+\frac{1}{n})^n$ approaches 1, we have the **important limit**

$$\left(1+\frac{1}{n}\right)^n \longrightarrow e^1 = e.$$

Now consider the following four sequences, which we will show represent different orders of growth as $n \longrightarrow \infty$:

$$\{(\log n)^k\}, \quad \{n^p\}, \quad \{a^n\}, \quad \{n!\}; \tag{21.1}$$

here $k > 0$, $p > 0$, and $a > 1$ are fixed positive numbers. The notation $n!$, read n-**factorial**, denotes the product of the first n positive integers:

$$n! = 1 \cdot 2 \cdot 3 \cdot \cdots \cdot n.$$

Each of the sequences (21.1) approaches ∞ as $n \longrightarrow \infty$, but they grow at very different rates, so that for any $k > 0$, $p > 0$, and $a > 1$,

$$\frac{(\log n)^k}{n^p} \longrightarrow 0; \quad \frac{n^p}{a^n} \longrightarrow 0; \quad \frac{a^n}{n!} \longrightarrow 0. \tag{21.2}$$

For example,

$$\frac{(\log n)^5}{\sqrt{n}} \longrightarrow 0; \quad \frac{n^{10}}{2^n} \longrightarrow 0; \quad \frac{1000^n}{n!} \longrightarrow 0.$$

We use l'Hospital's Rule to show that $\frac{\log x}{x^p} \longrightarrow 0$ if $p > 0$, and this of course shows that $\frac{\log n}{n^p} \longrightarrow 0$.

$$\lim_{x\to\infty} \frac{\log x}{x^p} = \lim_{x\to\infty} \frac{\frac{1}{x}}{px^{p-1}} = \lim_{x\to\infty} \frac{1}{p}\frac{1}{x^p} = 0.$$

We again use l'Hospital's Rule to show that $\frac{x^p}{a^x} \longrightarrow 0$ for $p = 1, 2, 3, \ldots$ and hence for any $p > 0$. If $a > 1$, then

$$\lim_{x\to\infty} \frac{x}{a^x} = \lim_{x\to\infty} \frac{1}{a^x \log a} = 0; \tag{21.3}$$

$$\lim_{x\to\infty} \frac{x^2}{a^x} = \lim_{x\to\infty} \frac{2x}{a^x \log a} = \frac{2}{\log a} \lim_{x\to\infty} \frac{x}{a^x} = 0; \tag{21.4}$$

$$\lim_{x\to\infty} \frac{x^3}{a^x} = \lim_{x\to\infty} \frac{3x^2}{a^x \log a} = \frac{3}{\log a} \lim_{x\to\infty} \frac{x^2}{a^x} = 0. \tag{21.5}$$

In equations (21.3)–(21.5) we used the result (21.3) in the last step of (21.4), and the result of (21.4) in the last step of (21.5). The process obviously can be continued to show that $\frac{x^n}{a^x} \longrightarrow 0$ for all positive integers n, and hence for arbitrarily large numbers p, and hence for any positive number p.

EXAMPLE 21.4

Show $\dfrac{(\log n)^3}{n^p} \longrightarrow 0$ for all $p > 0$.

Solution

We write $n^p = n^{\frac{p}{3}} \cdot n^{\frac{p}{3}} \cdot n^{\frac{p}{3}}$. Since $\frac{p}{3} > 0$, $\frac{(\log n)}{n^{\frac{p}{3}}} \longrightarrow 0$. Therefore,

$$\frac{(\log n)^3}{n^p} = \left(\frac{\log n}{n^{\frac{p}{3}}}\right)^3 \longrightarrow 0.$$

The argument of Example 21.4 will work for any power of $\log n$, so for any k and any $p > 0$,

$$\frac{(\log n)^k}{n^p} \longrightarrow 0.$$

EXAMPLE 21.5

Find $\lim\limits_{n\to\infty} \dfrac{(\log n)^2}{3^n}$.

Solution

Since 3^n dominates n^p and n^p dominates $(\log n)^2$ for any p, it follows that 3^n dominates $(\log n)^2$. We can make an explicit comparison as follows:

$$\frac{(\log n)^2}{3^n} = \left(\frac{\log n}{n}\right)^2 \frac{n^2}{3^n} \longrightarrow 0 \cdot 0 = 0.$$

EXAMPLE 21.6

Find $\lim\limits_{n\to\infty} \dfrac{n^4 \log n}{e^n}$.

Solution

Since $\log n < n$ for all large n,

$$\lim_{n\to\infty} \frac{n^4 \log n}{e^n} \leq \lim_{n\to\infty} \frac{n^5}{e^n} = 0.$$

The eyeball rule for fractional sequences like those in Examples 21.5 and 21.6 is to search out the dominant term – 3^n in Example 21.5 and e^n in Example 21.6. If the dominant term is in the denominator, the fraction tends to zero; if the dominant term is in the numerator, the fraction tends to ∞.

EXAMPLE 21.7

Find $\lim_{n\to\infty} \dfrac{n^{10}+\sqrt{n}\log n}{n!+n^2\log n}$.

Solution

The dominant term is $n!$, so the fraction tends to zero. We can make an explicit estimate by using $\sqrt{n} < n^9$ and $\log n < n$ to increase the numerator, and dropping the term $n^2 \log n$ to decrease the denominator. Thus,

$$\frac{n^{10}+\sqrt{n}\,\log n}{n!+n^2\log n} < \frac{n^{10}+n^9\cdot n}{n!} = 2\cdot\frac{n^{10}}{n!} \longrightarrow 0.$$

EXAMPLE 21.8

$\lim_{n\to\infty} n^{\frac{1}{n}}$.

This has the indeterminate form ∞^0. For any such indeterminate exponential expression, we first take the logarithm

$$\lim_{n\to\infty} \log n^{\frac{1}{n}} = \lim_{n\to\infty} \frac{1}{n}\log n = 0.$$

Since the logarithm tends to zero, the function tends to $e^0 = 1$; that is,

$$\lim_{n\to\infty} n^{\frac{1}{n}} = 1.$$

To see why $\frac{a^n}{n!} \longrightarrow 0$, no matter how large a is, let N be some number larger than $2a$, so $\frac{a}{N} < \frac{1}{2}$. Then for $n = N + k$,

$$\begin{aligned}\frac{a^n}{n!} &= \left[\frac{a\cdot a\cdot a\cdot\ \cdots\ \cdot a}{1\cdot 2\cdot 3\cdot\ \cdots\ \cdot N}\right]\cdot\left(\frac{a}{N+1}\right)\left(\frac{a}{N+2}\right)\cdots\left(\frac{a}{N+k}\right)\\ &< \left[\frac{a^N}{N!}\right]\left(\frac{1}{2}\right)^k.\end{aligned}$$

Since N is fixed, $\frac{a^N}{N!}$ is a fixed number, and $(\frac{1}{2})^k \longrightarrow 0$, $\frac{a^n}{n!} \longrightarrow 0$ as $n = N + k \longrightarrow \infty$.

In our later works we will be concerned with functions of the form $a_n x^n$. For these functions, the critical question will be for what values of x is $\lim_{n\to\infty} a_n x^n = 0$, and for what values of x is $\lim_{n\to\infty} a_n x^n = \infty$.

EXAMPLE 21.9

Find the values of x such that $\frac{n^2|x|^n}{2^n \log n} \longrightarrow 0$, and the values of x for which the limit is ∞.

Solution

The dominant terms are the exponentials $|x|^n$ and 2^n, so we write

$$\frac{n^2|x|^n}{2^n\log n} = \frac{n^2}{\log n}\left|\frac{x}{2}\right|^n.$$

If $|\frac{x}{2}| < 1$, the terms approach zero since the sequence is the same as $\frac{n^2}{(\log n)a^n}$ with $a = |\frac{2}{x}| > 1$. If $|\frac{x}{2}| > 1$, the terms tend to ∞, since $\frac{n^2 a^n}{\log n} \longrightarrow \infty$ if $a \geq 1$.

The same sort of argument shows that

$$\frac{(\log n)}{n^2}\left|\frac{x}{2}\right|^n \longrightarrow 0 \text{ if} |x| \leq 2,$$

$$\frac{(\log n)}{n^2}\left|\frac{x}{2}\right|^n \longrightarrow \infty \text{ if } |x| > 2.$$

We finish with a computation that shows how l'Hospital's Rule for $\frac{0}{0}$ as $x \longrightarrow \infty$ follows from l'Hospital's Rule for $\frac{0}{0}$ as $x \longrightarrow 0+$:

$$\begin{aligned}\lim_{x\to\infty} \frac{f(x)}{g(x)} &= \lim_{x\to 0+} \frac{f\left(\frac{1}{x}\right)}{g\left(\frac{1}{x}\right)} \\ &= \lim_{x\to 0+} \frac{f'\left(\frac{1}{x}\right)\left(-\frac{1}{x^2}\right)}{g'\left(\frac{1}{x}\right)\left(-\frac{1}{x^2}\right)} \\ &= \lim_{x\to 0+} \frac{f'\left(\frac{1}{x}\right)}{g'\left(\frac{1}{x}\right)} \\ &= \lim_{x\to\infty} \frac{f'(x)}{g'(x)}.\end{aligned}$$

PROBLEMS

21.1 $\lim_{x\to\infty} \frac{x^3}{1-x^3}$

21.2 $\lim_{x\to\infty} \frac{2x+1}{x^2+3}$

21.3 $\lim_{x\to\infty} 2^{-x}$

21.4 $\lim_{x\to\infty} \frac{\sin x}{5x+1}$

21.5 $\lim_{x\to\infty} \frac{x^4}{e^x}$

21.6 $\lim_{x\to\infty} \frac{(\log x)^2}{x}$

21.7 $\lim_{n\to\infty} \frac{n^2}{(1.02)^n}$

21.8 $\lim_{n\to\infty} \frac{\sqrt{n}}{n!}$

21.9 $\lim_{n\to\infty} \frac{10^n}{n!}$

21.10 $\lim_{n\to\infty} \frac{n\log n}{2^n}$

21.11 $\lim_{n\to\infty} n\sin\frac{1}{n}$ *Hint:* Consider $\frac{\sin x}{x}$ as $x \longrightarrow 0$.

21.12 $\lim_{n\to\infty} 2^{\frac{1}{n}}$

21.13 $\lim_{n\to\infty} n^2 e^{-n}$

21.14 $\lim_{n\to\infty} \frac{2^n \cdot 3^n}{5^n}$

21.15 $\lim_{n\to\infty} \frac{n^4}{4^n}$

21.16 $\lim_{n\to\infty} (1+\frac{2}{n})^n$

21.17 $\lim_{n\to\infty} \frac{(\log n)^{10} n^{100} 100^n}{n!}$

21.18 $\lim_{n\to\infty} \frac{n!}{e^n}$

21.19 $\lim_{n\to\infty} \frac{2^{n+1} \cdot n^2}{3^n}$

21.20 $\lim_{n\to\infty} \frac{\log n + n^2}{(\frac{1}{2})^n + \sqrt{n}}$

Find the values of x for which the following sequences approach zero, and the values of x for which they approach ∞.

21.21 $\frac{1}{n}|x|^n$

21.22 $n^3|x|^n$

21.23 $\frac{(\log n)|x|^n}{n^2}$

21.24 $\frac{2^n|x|^n}{n^3}$

21.25 $\frac{(n^2+1)|x|^n}{3^n+n} = \frac{(n^2+1)}{1+\frac{n}{3^n}}\left|\frac{x}{3}\right|^n$

21.26 $n!|x|^n$

21.27 $\frac{n!|x|^n}{(n+2)!} = \frac{|x|^n}{(n+2)(n+1)}$

21.28 $\frac{(n+2)!|x|^n}{n!}$

22

Improper Integrals

So far all our integrals $\int_a^b f(x)\,dx$ involve bounded functions $f(x)$ on bounded intervals $[a, b]$. Now we extend the definition to include some unbounded functions and some unbounded intervals. Such integrals are called **improper integrals**. The integral

$$\int_0^4 \frac{1}{\sqrt{x}}\,dx \tag{22.1}$$

is an example of an improper integral with an unbounded integrand, and

$$\int_1^\infty \frac{1}{x^2}\,dx \tag{22.2}$$

is an example of an improper integral over an unbounded interval.

For integrals over unbounded intervals (i.e., intervals of the form $[a, \infty)$ or $(-\infty, b]$), we make the definition:

$$\int_a^\infty f(x)\,dx = \lim_{b\to\infty} \int_a^b f(x)\,dx, \tag{22.3}$$

$$\int_{-\infty}^b f(x)\,dx = \lim_{a\to-\infty} \int_a^b f(x)\,dx. \tag{22.4}$$

Here we assume that $f(x)$ is integrable over every bounded subinterval of the interval of integration, so all the integrals on the right in (22.3) and (22.4) make sense. If the limit on the right in (22.3) or (22.4) exists, we say the improper integral on the left **converges**.

Consider, for example, the improper integral (22.2); by definition,

$$\begin{aligned}
\int_1^\infty \frac{1}{x^2}\,dx &= \lim_{b\to\infty} \int_1^b \frac{1}{x^2}\,dx \\
&= \lim_{b\to\infty} -\frac{1}{x}\Big]_1^b \\
&= \lim_{b\to\infty} \left[-\frac{1}{b} + \frac{1}{1}\right] = 1.
\end{aligned}$$

Since the limit exists, the improper integral $\int_1^\infty \frac{1}{x^2}\,dx$ converges, and its value is 1.

If the integrand $f(x)$ is unbounded at one end of an interval $[a, b]$, we again use a limit to define $\int_a^b f(x)\,dx$. If $f(x)$ is unbounded at b, but $\int_a^c f(x)\,dx$ exists for every c with $a < c < b$, then we define

$$\int_a^b f(x)\,dx = \lim_{c\to b-} \int_a^c f(x)\,dx.$$

The notation $c \longrightarrow b-$ means that c approaches b from the left side. If $f(x)$ is unbounded at a but integrable on $[c, b]$ for every c between a and b, then we define

$$\int_a^b f(x)\,dx = \lim_{c\to a+} \int_c^b f(x)\,dx,$$

where the limit is taken as c approaches a from the right side of a. If the limit that defines the integral exists, we say the integral **converges**; otherwise, the integral **diverges**, and we attach no meaning to it.

As an example of this second type of improper integral, consider the integral

$$\int_0^4 \frac{1}{\sqrt{x}}\,dx,$$

that is improper since $\frac{1}{\sqrt{x}} \longrightarrow \infty$ as $x \longrightarrow 0+$. From the definition we have

$$\begin{aligned}\int_0^4 \frac{1}{\sqrt{x}}\,dx &= \lim_{c\to 0+} \int_c^4 \frac{1}{\sqrt{x}}\,dx \\ &= \lim_{c\to 0+} 2\sqrt{x}\Big]_c^4 \\ &= \lim_{c\to 0+} (2\sqrt{4} - 2\sqrt{c}) = 4.\end{aligned}$$

The most significant improper integrals are those with positive integrands, and we will stick to that case. If $f(x) \geq 0$, then $\int_a^b f(x)\,dx$ or $\int_a^\infty f(x)\,dx$ represents the area of the region under the curve $y = f(x)$, and convergence of the integral means that this area is finite even though the region is unbounded. It is clear from this area interpretation that the following comparison theorem holds: If $0 \leq f(x) \leq kg(x)$ *for* $x \geq a$ *and* $\int_a^\infty g(x)\,dx$ *converges* (so that $\int_a^\infty kg(x)\,dx$ *converges*), *then* $\int_a^\infty f(x)\,dx$ *converges*. In other words, if there is finite area under $y = kg(x)$, then there is finite area under the lower curve $y = f(x)$. A similar statement, of course, holds for improper integrals like $\int_1^2 \frac{dx}{2-x}$ or $\int_0^1 \frac{dx}{x^{\frac{1}{3}}}$ where the integrand is unbounded at one end of the interval.

We will pay particular attention to integrals of the form $\int_a^\infty f(x)\,dx$, with $f(x) \geq 0$, since these integrals figure importantly in the study of infinite series.

EXAMPLE 22.1

(a) $\displaystyle\int_1^\infty \frac{1}{\sqrt{x}}\,dx$; (b) $\displaystyle\int_1^\infty \frac{1}{x^2}\,dx$.

Using the definition, we calculate as follows:

$$\begin{aligned}\int_1^\infty \frac{1}{\sqrt{x}}\,dx &= \lim_{b\to\infty} \int_1^b \frac{1}{\sqrt{x}}\,dx \\ &= \lim_{b\to\infty} 2\sqrt{x}\Big]_1^b \\ &= \lim_{b\to\infty} (2\sqrt{b} - 2) = \infty.\end{aligned}$$

The limit does not exist, and the integral diverges.

The calculation for (b) is similar, but this time the integral converges.

$$\int_1^\infty \frac{1}{x^2}\,dx = \lim_{b\to\infty}\int_1^b \frac{1}{x^2}\,dx$$
$$= \lim_{b\to\infty} -\frac{1}{x}\Big]_1^b$$
$$= \lim_{b\to\infty}\left(-\frac{1}{b}+1\right) = 1.$$

The limit exists, so the integral converges and has the value one (1):

$$\int_1^\infty \frac{1}{x^2}\,dx = 1.$$

Example 22.1 shows that $\int_1^\infty \frac{1}{x^p}\,dx$ converges for some p (e.g., $p = 2$) and diverges for some p (e.g., $p = \frac{1}{2}$). Since $\frac{1}{x^p}$ gets smaller on $[1, \infty)$ as p gets bigger, there must be a critical point p_0 such that the integral $\int_1^\infty \frac{1}{x^p}\,dx$ converges if $p > p_0$ and diverges if $p < p_0$. Here is the calculation:

$$\int_1^\infty \frac{1}{x^p}\,dx = \lim_{b\to\infty}\int_1^b x^{-p}\,dx$$
$$= \lim_{b\to\infty} \frac{1}{-p+1}x^{-p+1}\Big]_1^b$$
$$= \lim_{b\to\infty} \frac{1}{-p+1}[b^{-p+1}-1].$$

If $p > 1$, then the exponent in b^{1-p} is negative, $\lim_{b\to\infty} b^{1-p} = 0$, and the integral converges. If $p < 1$, then $1 - p > 0$ and $\lim_{b\to\infty} b^{1-p} = \infty$. Therefore, $\int_1^\infty \frac{1}{x^p}\,dx$ converges if $p > 1$ and diverges if $p < 1$. If $p = 1$, the integration formula is different and we check that case separately:

$$\int_1^\infty \frac{1}{x}\,dx = \lim_{b\to\infty}\int_1^b \frac{1}{x}\,dx$$
$$= \lim_{b\to\infty} \log x\Big]_1^b$$
$$= \lim_{b\to\infty}(\log b - 0) = \infty.$$

The integral diverges if $p = 1$. Hence

$$\int_1^\infty \frac{1}{x^p}\,dx \textit{ converges if and only if } p > 1.$$

The lower limit in the integrals above was taken to be 1 for simplicity. Of course, the lower limit must be positive since $\frac{1}{x^p}$ is unbounded at 0. However, for any $a > 0$, $\int_a^\infty f(x)\,dx$ converges if and only if $\int_1^\infty f(x)\,dx$ converges, since the area between $x = 1$ and $x = a$ is certainly finite.

Integrals over unbounded intervals $(-\infty, b]$ are treated in exactly the same way:

$$\int_{-\infty}^b f(x)\,dx = \lim_{a\to-\infty}\int_a^b f(x)\,dx.$$

Now consider integrals of the type

$$\int_0^a \frac{1}{x^p}\,dx, \tag{22.5}$$

where $a > 0$ and $p > 0$. The convergence or divergence of (22.5) does not depend on how big a is, so we let $a = 1$ for convenience. Consider the following two special cases:

$$(a) \int_0^1 \frac{1}{\sqrt{x}}\,dx; \quad (b) \int_0^1 \frac{1}{x^2}\,dx.$$

The same simple integration as in Example 22.1 shows that

$$\int_0^1 \frac{1}{\sqrt{x}}\,dx = \lim_{a\to 0+} \int_a^1 \frac{1}{\sqrt{x}}\,dx = \lim_{x\to 0+} (2 - 2\sqrt{a}) = 2,$$

$$\int_0^1 \frac{1}{x^2}\,dx = \lim_{a\to 0+} \int_a^1 \frac{1}{x^2}\,dx = \lim_{a\to 0+} \left(-1 + \frac{1}{a}\right] = \infty.$$

The functions $\frac{1}{x^p}$ for $p > 0$ are all unbounded at 0, and some integrals converge (e.g., $p = \frac{1}{2}$) while others diverge (e.g., $p = 2$). The bigger p is, the faster $\frac{1}{x^p}$ grows as $x \longrightarrow 0+$, and the same kind of calculation as we made for integrals $\int_1^\infty \frac{1}{x^p}\,dx$ shows that

$$\int_0^1 \frac{1}{x^p}\,dx \textit{ converges if and only if } p < 1.$$

There is an obvious similarity between integrals $\int_0^1 \frac{1}{x^p}\,dx$ and integrals $\int_1^\infty \frac{1}{x^p}\,dx$. For positive p, $p > 1$ if and only if $\frac{1}{p} < 1$, so

$$\int_1^\infty \frac{1}{x^p}\,dx \quad \text{and} \quad \int_0^1 \frac{1}{x^{\frac{1}{p}}}\,dx$$

both converge or both diverge. The geometry makes the situation even clearer. Consider, for example, the functions $\frac{1}{\sqrt{x}}$ and $\frac{1}{x^2}$. These functions are inverses of each other, since

$$\frac{1}{\sqrt{\frac{1}{x^2}}} = \frac{1}{\left(\frac{1}{\sqrt{x}}\right)^2} = x,$$

so their graphs are symmetric about the line $y = x$. Thus, the two shaded areas in Figure 22.1 are equal; that is,

$$\int_0^1 \left(\frac{1}{\sqrt{x}} - 1\right) dx = \int_1^\infty \frac{1}{x^2}\,dx = 1.$$

Similarly, the curves $y = \frac{1}{x^3}$ and $y = \frac{1}{x^{\frac{1}{3}}}$ are symmetric about the line $y = x$, and both of the following integrals converge:

$$\int_0^1 \left(\frac{1}{\sqrt[3]{x}} - 1\right) dx = \int_1^\infty \frac{1}{x^3}\,dx = \frac{1}{2}.$$

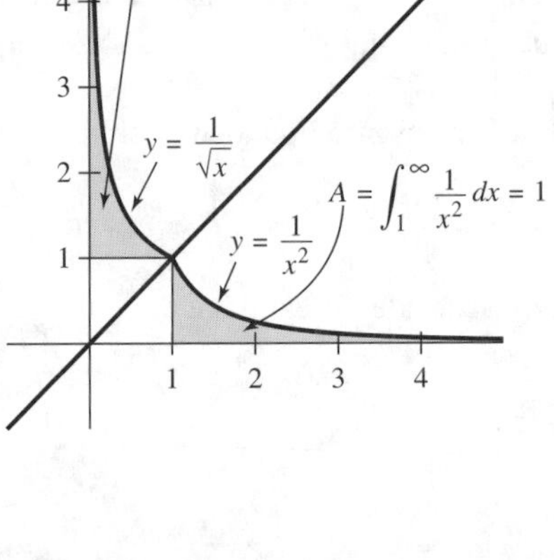

Figure 22.1

EXAMPLE 22.2

(a) $\displaystyle\int_0^1 \frac{\sin x}{x^{\frac{3}{2}}}\,dx.$

For $0 \le x \le 1$, $0 \le \sin x \le x$. Therefore,

$$\frac{\sin x}{x^{\frac{3}{2}}} \le \frac{x}{x^{\frac{3}{2}}} = \frac{1}{\sqrt{x}}.$$

Since $\int_0^1 \frac{1}{\sqrt{x}}\,dx$ converges ($p = \frac{1}{2} < 1$), the given integral with a smaller integrand also converges.

EXAMPLE 22.3

$\displaystyle\int_e^\infty \frac{x}{1+x^3}\,dx.$

First notice that the convergence or divergence of the integral doesn't depend on the lower limit. Since $\frac{x}{(1+x^3)}$ behaves roughly like $\frac{x}{x^3} = \frac{1}{x^2}$ for large x, we make this comparison:

$$\frac{x}{1+x^3} < \frac{x}{x^3} = \frac{1}{x^2}.$$

Since $\int_e^\infty \frac{1}{x^2}\,dx$ converges ($p = 2 > 1$), the given smaller integral converges.

EXAMPLE 22.4

$\displaystyle\int_0^\infty \frac{e^{-x}}{\sqrt{x}}\,dx.$

This integral is improper for two reasons—the integrand is unbounded at 0, and the interval $[0, \infty)$ is unbounded. All such integrals must be broken up into integrals with a single impropriety, and convergence of the integral requires convergence of all the pieces. Thus, we write

$$\int_0^\infty \frac{e^{-x}}{\sqrt{x}}\,dx = \int_0^1 \frac{e^{-x}}{\sqrt{x}}\,dx + \int_1^\infty \frac{e^{-x}}{\sqrt{x}}\,dx,$$

and check the two integrals separately for convergence. For $0 \le x \le 1$, $e^{-x} \le 1$, so $\frac{e^{-x}}{\sqrt{x}} \le \frac{1}{\sqrt{x}}$, and the first integral converges. For $x \ge 1$, $\frac{e^{-x}}{\sqrt{x}} \le e^{-x}$, and $\int_1^\infty e^{-x}\,dx$ converges:

$$\begin{aligned}\int_1^\infty e^{-x}\,dx &= \lim_{b\to\infty}\int_1^b e^{-x}\,dx \\ &= \lim_{b\to\infty}\,[-e^{-x}]_1^b \\ &= \lim_{b\to\infty}\left(e^{-b} + e^{-1}\right) = \frac{1}{e}.\end{aligned}$$

Since $\int_1^\infty e^{-x}\,dx$ converges, $\int_1^\infty \frac{e^{-x}}{\sqrt{x}}\,dx$ also converges. Therefore, $\int_0^\infty \frac{e^{-x}}{\sqrt{x}}\,dx$ converges.

PROBLEMS

Evaluate the integral, or show that it diverges.

22.1 $\displaystyle\int_0^\infty e^{-x}\,dx$

22.2 $\displaystyle\int_8^\infty \frac{dx}{x^{\frac{5}{3}}}$

22.3 $\displaystyle\int_0^\infty \frac{dx}{1+x^2}$

22.4 $\int_e^\infty \frac{dx}{x \log x}$

22.5 $\int_0^\infty xe^{-x}\,dx$

22.6 $\int_2^\infty \frac{dx}{x^2}$

22.7 $\int_1^\infty \frac{dx}{x^{\frac{2}{3}}}$

22.8 $\int_1^5 \frac{dx}{x-1}$

22.9 $\int_0^1 \frac{dx}{1-x^2}$

22.10 $\int_0^1 \frac{dx}{\sqrt{1-x}}$

22.11 $\int_1^\infty \frac{\log x}{x}\,dx$

22.12 $\int_0^1 \frac{dx}{\sqrt{1-x^2}}$

22.13 $\int_e^\infty \frac{\log x}{x^2}\,dx$ (Integrate by parts with $u = \log x$.)

22.14 $\int_0^\infty \frac{dx}{9+4x^2}$

22.15 $\int_1^\infty \frac{dx}{x^2-x}$

22.16 $\int_1^\infty (\frac{1}{\sqrt{x+1}} - \frac{1}{\sqrt{x}})\,dx.$ *Hint:* $\sqrt{x+1} - \sqrt{x} = \frac{1}{\sqrt{x+1}+\sqrt{x}}$.

Use the comparison test to tell whether the following converge or diverge.

22.17 $\int_1^\infty \frac{dx}{\sqrt{1+x^4}}$

22.18 $\int_0^\infty \frac{e^{-x}}{1+x^2}\,dx$

22.19 $\int_0^\infty (1+x^3)^{-4}\,dx$

22.20 $\int_1^\infty \frac{1+\cos x}{x^2}\,dx$

Find the volume obtained by rotating about the x-axis the following areas:

22.21 The area under $y = \frac{1}{x}$ for $1 \le x < \infty$.

22.22 The area under $y = e^{-x}$ for $1 \le x < \infty$.

22.23 The area under $y = \frac{1}{\sqrt{1+x^2}}$ for $0 \le x < \infty$.

22.24 The area under $y = xe^{-x}$ for $0 \le x < \infty$.

22.25 (i) Show that $\int_0^\infty xe^{-x}\,dx = 1$.

(ii) Use integration by parts with $u = x^{n+1}$ to show that $\int_0^\infty x^{n+1}e^{-x}\,dx = (n+1)\int_0^\infty x^n e^{-x}\,dx$.

(iii) Use (i) and (ii) to find $\int_0^\infty x^n e^{-x}\,dx$ for $n = 2, 3, 4, \ldots$.

23

Series

We extend the operation of addition from a finite number of terms to an infinite number of terms by taking a limit; that is, we define the infinite sum

$$a_1 + a_2 + a_3 + \cdots + a_n + \cdots \tag{23.1}$$

to be a limit of the finite sums s_n, where

$$s_n = a_1 + a_2 + \cdots + a_n. \tag{23.2}$$

Any indicated infinite sum like (1) is called a **series**, and the sum of the series (23.1) is the limit of the *sequence* $\{s_n\}$:

$$a_1 + a_2 + \cdots + a_n + \cdots = \lim_{n\to\infty} s_n. \tag{23.3}$$

The numbers $\{s_n\}$, with each s_n defined by (23.2), form the **sequence of partial sums** of the series (1). The number s_n is the n**th partial sum** of the series.

The following notation for series is convenient:

$$\sum_{n=1}^{\infty} a_n = a_1 + a_2 + \cdots + a_n + \cdots .$$

This $\sum$-notation can also be used for finite sums. For example,

$$\sum_{n=1}^{5} a_n = a_1 + a_2 + a_3 + a_4 + a_5,$$

$$\sum_{n=1}^{6} 2 = 2 + 2 + 2 + 2 + 2 + 2 = 12,$$

$$\sum_{k=1}^{n} a_k = a_1 + a_2 + \cdots + a_n = s_n.$$

We say the series $\sum_{n=1}^{\infty} a_n$ **converges** if the *sequence* $\{s_n\}$ of its partial sums converges. If the sequence $\{s_n\}$ does not converge, the series **diverges**. We will frequently omit the range on the indices when dealing with infinite series, and write $\sum a_n$ instead of $\sum_{n=1}^{\infty} a_n$.

If $\sum a_n$ is a convergent series, with $\{s_n\}$ its sequence of partial sums, then $s_n \longrightarrow s$, and $s_{n-1} \longrightarrow s$, so

$$a_n = s_n - s_{n-1} \longrightarrow s - s = 0.$$

A series cannot converge unless the terms tend to zero. The condition $a_n \longrightarrow 0$ is NOT sufficient for convergence, only necessary. We will see many divergent series whose terms tend to zero.

EXAMPLE 23.1

Show that the following series diverge.

$$(a)\ \sum \frac{n}{2n+1}; \quad (b)\ \sum (-1)^n.$$

In series (a), $a_n = \frac{n}{2n+1} \longrightarrow \frac{1}{2}$. The terms approach a limit, but the limit is not zero, so the series diverges. In (b), the terms $(-1)^n$ oscillate between 1 and -1, so the sequence $\{a_n\}$ does not approach a limit, and the series diverges.

It is easy to see that if $\sum a_n$ converges, then $\sum c a_n$ also converges and

$$\sum c a_n = c \sum a_n.$$

Similarly, you can add the corresponding terms of two convergent series, so that if $\sum a_n$ and $\sum b_n$ both converge, then $\sum (a_n + b_n)$ converges and

$$\sum (a_n + b_n) = \sum a_n + \sum b_n.$$

The convergence or divergence of a series has nothing to do with the first 100 terms, or the first 100 million terms. It is only the tail of the series that determines convergence. Therefore, to determine convergence or divergence, we need only consider how the terms a_n behave for all sufficiently large n.

The following two series are instructive:

$$1 + \frac{1}{2} + \frac{1}{3} + \frac{1}{4} + \cdots + \frac{1}{n} + \cdots, \tag{23.4}$$

$$1 - \frac{1}{2} + \frac{1}{3} - \frac{1}{4} + \cdots \pm \frac{1}{n} \mp \cdots. \tag{23.5}$$

The first series, called the **harmonic series**, diverges, and the second series converges. To see that the harmonic series (23.4) diverges, consider the following partial sums:

$$s_1 = 1 \geq \frac{1}{2},$$

$$s_2 = 1 + \frac{1}{2} \geq 2 \cdot \frac{1}{2},$$

$$s_4 = 1 + \frac{1}{2} + \left(\frac{1}{3} + \frac{1}{4}\right) \geq 3 \cdot \frac{1}{2},$$

$$s_8 = 1 + \frac{1}{2} + \left(\frac{1}{3} + \frac{1}{4}\right) + \left(\frac{1}{5} + \frac{1}{6} + \frac{1}{7} + \frac{1}{8}\right) \geq 4 \cdot \frac{1}{2}.$$

Continuing this way, we see that $s_{16} \geq 5 \cdot \frac{1}{2}$, $s_{32} \geq 6 \cdot \frac{1}{2}$, and so on. Clearly, $s_n \longrightarrow \infty$, so $\sum \frac{1}{n}$ diverges.

Now consider (23.5), and more generally any series of the form

$$a_1 - a_2 + a_3 - a_4 + \cdots, \tag{23.6}$$

where $a_1 \geq a_2 \geq a_3 \geq \cdots$ and $a_n \longrightarrow 0$. The partial sums s_n of (23.6) start at $s_1 = a_1$, and then jump successively to the left and right as a_2 is subtracted, a_3 is added, a_4 subtracted, and so on. The jumps get smaller since the a_n decrease, and all partial sums beyond s_n lie in a fixed interval of length a_n. These intervals collapse to a single point since $a_n \longrightarrow 0$, so $\{s_n\}$ converges to that point. *Any series converges if the signs alternate, the terms decrease in magnitude, and the terms tend to zero.* We will call such a series a **proper alternating series**, where the word "proper" indicates that not only do the signs alternate, but the other two conditions are also satisfied, so *any proper alternating series converges.* Moreover, because of the way the s_n jump back and forth in a proper alternating series, it is clear that s_n is always within a distance a_{n+1} of the limiting sum.

EXAMPLE 23.2

The following series is a proper alternating series, and it is known that the sum is $\frac{\pi}{4}$:

$$1 - \frac{1}{3} + \frac{1}{5} - \frac{1}{7} + \cdots = \frac{\pi}{4}.$$

How many terms must you add to get an approximation to $\frac{\pi}{4}$ accurate to within .05?

We know the error between s_n in this alternating series and the sum, $\frac{\pi}{4}$, is less than the first term omitted. If the first term omitted is $\frac{1}{21}$, the error will be less than .05. Your calculator will show

$$1 - \frac{1}{3} + \frac{1}{5} - \frac{1}{7} + \frac{1}{9} - \frac{1}{11} + \frac{1}{13} - \frac{1}{15} + \frac{1}{17} - \frac{1}{19} \doteq .760.$$

Your calculator will also show that $\frac{\pi}{4} \doteq .785$, so .760 is indeed accurate within .05. Notice that since the partial sums jump back and forth over the limit, the number halfway between the sum to $-\frac{1}{19}$ and the sum to $+\frac{1}{21}$ is a much better approximation; that is,

$$\left(1 - \frac{1}{3} + \frac{1}{5} - \cdots - \frac{1}{19}\right) + \frac{1}{2} \cdot \frac{1}{21} \doteq .760 + .024 = .784.$$

EXAMPLE 23.3

Tell whether the series converges or diverges, and why:

$$\text{(a)} \sum (-1)^n \frac{(\log n)}{n}; \quad \text{(b)} \sum (-1)^n \frac{n^2}{3n^2 + 1}.$$

The first series, (a), converges because the signs alternate, and $\frac{(\log n)}{n}$ decreases, and decreases to zero. The series (a) is therefore a proper alternating series. The series (b) diverges because $\frac{n^2}{(3n^2+1)} \nrightarrow 0$. Don't be misled by alternating signs. The terms must decrease to zero or it isn't a proper alternating series.

A series $\sum a_n$ such that $\sum a_n$ converges but $\sum |a_n|$ diverges is called **conditionally convergent**. The series (23.5) is conditionally convergent. If $\sum a_n$ and $\sum |a_n|$ both converge, then $\sum a_n$ is **absolutely convergent**. It is a theorem that if the series $\sum |a_n|$ of absolute values converges, then the series $\sum a_n$ necessarily also converges. Any cancellation because of differing signs of the a_n only helps the convergence.

A very simple and very important series is the **geometric series**

$$\sum_{n=0}^{\infty} ax^n = a + ax + ax^2 + \cdots + ax^n + \cdots. \tag{23.7}$$

If $x \neq 1$, we find a formula for s_n as follows:

$$\begin{aligned} s_n &= a + ax + ax^2 + \cdots + ax^n, \\ xs_n &= ax + ax^2 + \cdots + ax^n + ax^{n+1}, \\ (1-x)s_n &= a - ax^{n+1}, \\ s_n &= \frac{a - ax^{n+1}}{1-x}. \end{aligned} \tag{23.8}$$

The formula (23.8) works for any geometric series with $x \neq 1$, but the series converges only if $|x| < 1$. If $|x| < 1$, then $x^{n+1} \longrightarrow 0$, so $\frac{ax^{n+1}}{(1-x)} \longrightarrow 0$, and $s_n \longrightarrow \frac{a}{(1-x)}$. Hence, for $-1 < x < 1$,

$$\sum_{n=0}^{\infty} ax^n = a + ax + ax^2 + \cdots + ax^n + \cdots = \frac{a}{1-x}.$$

EXAMPLE 23.4

Find the sums of the geometric series.

$$(a) \quad \sum_{n=1}^{\infty} \frac{1}{2^n}; \quad (b) \quad 0.333\ldots .$$

(a) The common ratio is $\frac{1}{2} < 1$, so the series converges. The first term is $a = \frac{1}{2}$, so

$$\sum_{n=1}^{\infty} \frac{1}{2^n} = \frac{\frac{1}{2}}{1 - \frac{1}{2}} = 1.$$

(b) Repeating decimals are simply a way of indicating a convergent geometric series. Here we have

$$\begin{aligned} 0.333\ldots &= \frac{3}{10} + \frac{3}{100} + \frac{3}{1000} + \cdots \\ &= \frac{\frac{3}{10}}{1 - \frac{1}{10}} = \frac{\frac{3}{10}}{\frac{9}{10}} = \frac{1}{3}. \end{aligned}$$

A conditionally convergent series has both positive and negative terms; the positive terms add up to $+\infty$, and the negative terms to $-\infty$. Conditional convergence therefore depends on a delicate cancellation between the positive and negative terms. It can be shown that a conditionally convergent series can be rearranged to converge to anything you like, or to diverge. Since the order in which the terms are added is all important in a conditionally convergent series, this kind of summation is not an entirely satisfactory generalization of a finite sum. On the other hand, if a series converges absolutely, then any rearrangement will also converge, and converge to the same number. Moreover, if $\sum a_n$ and $\sum b_n$ both converge absolutely, then you can add up all the products $a_n b_k$, in any order, and the result will be the product of $(\sum a_n)$ and $(\sum b_k)$ as it ought to be. No such statement can be made about conditionally convergent series. Absolute convergence is the property that allows us to treat infinite series pretty much like finite sums in terms of rearrangement, grouping, and multiplying.

Since checking a series $\sum a_n$ for absolute convergence involves checking the positive series $\sum |a_n|$, we now develop some tests for convergence of positive series; that is, series $\sum a_n$ with $a_n \geq 0$ for all n. If s_n is the nth partial sum of a positive series, then $\{s_n\}$ is an increasing sequence since each new term is obtained by adding a positive number; that is, $s_{n+1} = s_n + a_{n+1}$, and $a_{n+1} \geq 0$. It is a basic property of numbers that an increasing sequence is either bounded, and converges, or is unbounded and diverges to $+\infty$. Therefore, a positive series converges if and only if its partial sums remain bounded. This observation immediately gives us the following comparison test:

Comparison Test: *If* $0 \le a_n \le b_n$ *and* $\sum b_n$ *converges, then* $\sum a_n$ *converges; if* $\sum a_n$ *diverges, then* $\sum b_n$ *diverges.*

EXAMPLE 23.5

(a) $\sum \frac{3n}{n+1}\left(\frac{2}{3}\right)^n$; (b) $\sum \frac{\log n}{n}$.

(a) This series converges by comparison with the geometric series $\sum 3 \cdot (\frac{2}{3})^n$, since for all n,

$$\frac{3n}{n+1}\left(\frac{2}{3}\right)^n \le 3 \cdot \left(\frac{2}{3}\right)^n.$$

(b) Since $\sum \frac{1}{n}$ diverges and $\frac{\log n}{n} \ge \frac{1}{n}$ for $n \ge 3$, the series $\sum \frac{\log n}{n}$ diverges.

Our next test makes a comparison between the partial sums s_n of a positive series, and integrals $\int_1^n f(x)\,dx$ of a positive function. The improper integral $\int_1^\infty f(x)\,dx$ converges if and only if the integrals $\int_1^n f(x)\,dx$ remain bounded. We see from Figure 23.1 that if f is positive and decreasing,

$$f(1) + f(2) + \cdots + f(n-1) \ge \int_1^n f(x)\,dx, \tag{23.9}$$

and

$$f(2) + f(3) + \cdots + f(n) \le \int_1^n f(x)\,dx. \tag{23.10}$$

The first inequality shows that if $\sum f(n)$ converges, then $\int_1^\infty f(x)\,dx$ converges. The second inequality shows that if $\int_1^\infty f(x)\,dx$ converges, then the series $\sum f(n)$ converges. Therefore,

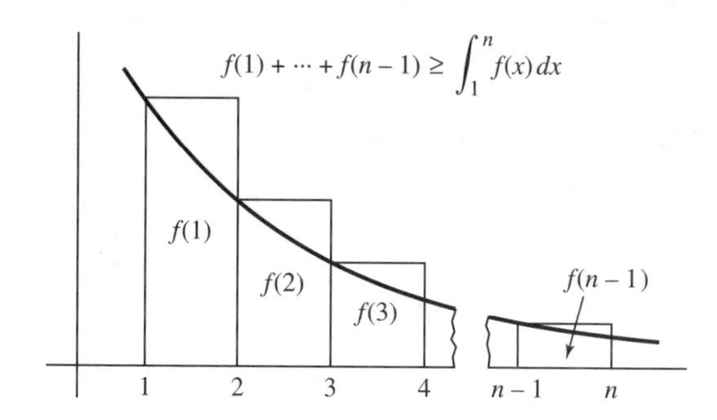

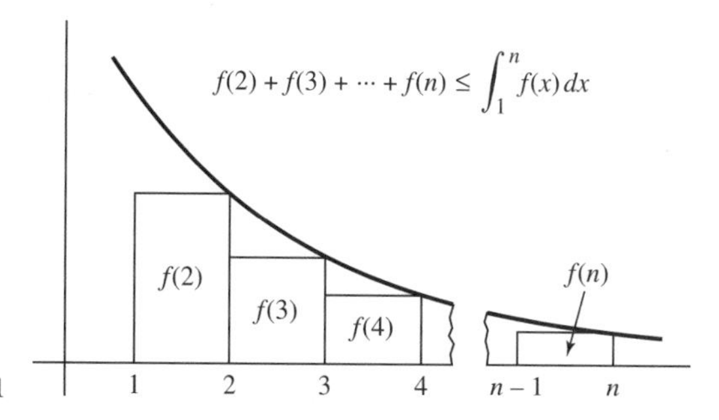

Figure 23.1

we have the following convergence test:

Integral Test: *If $f(x)$ is a positive decreasing function, then $\sum f(n)$ converges if and only if $\int_1^\infty f(x)\,dx$ converges.*

EXAMPLE 23.6

$$\sum_{n=2}^{\infty} \frac{1}{n \log n}.$$

The function $f(x) = \frac{1}{x \log x}$ is positive and decreasing for $x \geq 2$. We can therefore compare the series with $\int_2^\infty \frac{1}{x \log x}\,dx$.

The integrand has the form $\frac{1}{u}\,du$, with $u = \log x$, $du = \frac{dx}{x}$, so

$$\begin{aligned}\int_2^\infty \frac{1}{x \log x}\,dx &= \lim_{n\to\infty} \int_2^n \frac{1}{x \log x}\,dx \\ &= \lim_{n\to\infty} \log|\log x|\Big]_2^n \\ &= \lim_{n\to\infty} \log|\log n| - \log|\log 2| = \infty.\end{aligned}$$

Since the integral diverges, the series diverges.

Series of the form $\sum \frac{1}{n^p}$ are called p-**series**. If $p \leq 0$, $\frac{1}{n^p} \not\to 0$ and the series surely diverges, so we consider only p-series for $p > 0$. The divergent harmonic series is a p-series with $p = 1$. We saw in Chapter 22 that $\int_1^\infty \frac{1}{x^p}\,dx$ converges if and only if $p > 1$, so *a p-series converges if and only if $p > 1$.*

EXAMPLE 23.7

(a) $\sum \frac{1}{n\sqrt{n}}$; (b) $\sum \frac{n}{n^3+1}$; (c) $\sum \frac{2+\sin n}{\sqrt{n}}$.

(a) This is a p-series with $p = \frac{3}{2} > 1$, so the series converges.

(b) Clearly, $\frac{n}{(n^3+1)} \leq \frac{1}{n^2}$, and $\sum \frac{1}{n^2}$ is a convergent p-series . The given series therefore converges.

(c) The numerator $2 + \sin n$ is always greater than or equal to 1, so $\frac{(2+\sin n)}{\sqrt{n}} \geq \frac{1}{\sqrt{n}}$. Since $\sum \frac{1}{\sqrt{n}}$ is a divergent p-series ($p = \frac{1}{2} < 1$), the given series diverges.

An easy way to make the kind of comparison in (b) and (c) is given in the following test.

Limit Comparison Test: *If $\sum a_n$ and $\sum b_n$ are positive series, and* $\lim_{n\to\infty} \frac{a_n}{b_n} = \ell \neq 0$, *then $\sum a_n$ and $\sum b_n$ both converge, or both diverge.*

The limit comparison test works because if $\frac{a_n}{b_n} \longrightarrow \ell \neq 0$, then a_n and b_n are roughly multiples of each other for large n. Specifically, for all large n, $\frac{a_n}{b_n}$ must be close enough to ℓ so that $\frac{\ell}{2} \leq \frac{a_n}{b_n}$ and $\frac{a_n}{b_n} \leq 2\ell$. Hence, $b_n \leq \frac{2}{\ell} a_n$ and $a_n \leq 2\ell b_n$ for all large n. The first inequality shows that if $\sum a_n$ converges, so that $\sum \frac{2}{\ell} a_n$ converges, then $\sum b_n$ also converges. The second inequality shows that if $\sum b_n$ converges, then $\sum a_n$ converges.

If $\frac{a_n}{b_n} \longrightarrow 0$, then $a_n \leq b_n$ for all large n and consequently $\sum a_n$ converges if $\sum b_n$ converges, but not conversely. If $\frac{a_n}{b_n} \longrightarrow \infty$, then $b_n \leq a_n$ for all large n and $\sum b_n$ converges if $\sum a_n$ converges, but not conversely.

EXAMPLE 23.8

$$\sum \frac{n^2 + 100n}{3n^3 + 1}.$$

Here the dominant term in the numerator is n^2, and the dominant term in the denominator is $3n^3$. Therefore, for large n, the fraction will behave like $\frac{n^2}{3n^3} = \frac{1}{3n}$, so we make a limit comparison with $\sum \frac{1}{3n}$:

$$\lim_{n\to\infty} \frac{n^2 + 100n}{3n^3 + 1} \cdot \frac{3n}{1} = \lim_{n\to\infty} \frac{3n^3 + 100n^2}{3n^3 + 1} = 1.$$

The limit is nonzero, so both series behave in the same way. $\sum \frac{1}{3n}$ diverges so that the given series diverges.

PROBLEMS

Find the following finite sums. Brute force with a calculator will work, but a little thought will save time and be more interesting.

23.1 $\sum_{n=1}^{5} 100$

23.2 $\sum_{n=1}^{100} n.$ *Hint:* If $S = \sum_{n=1}^{100} n$, then also $S = \sum_{n=1}^{100} (101 - n)$. Hence $S + S = \sum_{n=1}^{100} 101$.

23.3 $\sum_{n=1}^{20} n$

23.4 $\sum_{n=1}^{10} (2n - 1) = 1 + 3 + 5 + \cdots + 19$

$$= (1 + 19) + (3 + 17) + (5 + 15) + (7 + 13) + (9 + 11)$$

23.5 $\sum_{n=1}^{N} (2n - 1) = 1 + 3 + 5 + \cdots + (2N - 1)$

23.6 $\sum_{n=1}^{10} \frac{2}{3^n}$

23.7 $\sum_{n=2}^{8} \left(\frac{3}{2}\right)^n$

Show that the series converges, or show that it diverges.

23.8 $\sum (-1)^n \frac{n}{n+1}$

23.9 $\sum (-1)^n \frac{1}{n^{\frac{3}{2}}}$

23.10 $\sum (-1)^n \frac{\log n}{n+1}$

23.11 $\sum (-1)^n \frac{n^3}{2^n}$

23.12 $\sum \frac{1}{\log n}$

23.13 $\sum \frac{\log n}{n}$

23.14 $\sum e^{-n}$

23.15 $\sum \frac{n}{e^n}$

23.16 $\sum \frac{e^n}{n^2}$

23.17 $\sum \frac{1}{(n+1)\log n}$

23.18 $\sum \frac{n}{n^2+1}$

23.19 $\sum \frac{1}{n} \sin \frac{1}{n}$. *Hint:* Compare with $\frac{1}{n^2}$.

23.20 $\sum \frac{4^n}{n \cdot 3^n}$. *Hint:* $\lim_{n\to\infty} \frac{4^n}{n3^n} = ?$

23.21 $\sum e^{\frac{1}{n}}$

23.22 $\sum \frac{2^n}{3^n + 4^n}$

23.23 Show that if $a_n \geq 0$ for all n and $\sum a_n$ converges, then $\sum a_n^2$ converges.

23.24 Show that if $a_n \geq 0$ for all n and $\sum a_n$ converges, then $\sum \sin a_n$ converges.

24

Power Series

Series of the form

$$\sum_{n=0}^{\infty} a_n x^n \quad \text{or} \quad \sum_{n=0}^{\infty} a_n (x - x_0)^n \tag{24.1}$$

are called **power series**. The sum of a power series is a function of x, and all our elementary functions ($\log x$, e^x, $\sin x$, $\sqrt{x}$, $\cos^{-1} x$, rational functions, etc.) can be expressed locally as power series. The two types of series in (24.1), one in powers of x, and one in powers of $(x - x_0)$, are obviously similar, and any statement about one type of series implies a corresponding statement about the other kind. For example, if $\sum a_n x^n$ converges for all x such that $|x| < 2$, then $\sum a_n (x-3)^n$ converges for all x such that $|x - 3| < 2$, and vice versa. We will therefore make our calculations with series in powers of x for simplicity; we realize that all our results will apply with appropriate changes to series in powers of $(x - x_0)$.

The power series $\sum a_n x^n$ converges or diverges depending on the value of x. Since the terms $|a_n x^n|$ get larger or smaller as $|x|$ gets larger or smaller, we expect that the closeness of x to the origin is the key to convergence. It is, and here is the precise statement:

Convergence Condition: *If* $\lim_{n\to\infty} a_n x_0^n = 0$ *then* $\sum a_n x^n$ *converges absolutely if* $|x| < |x_0|$. *If* $a_n x_0^n \nrightarrow 0$, *then* $\sum a_n x^n$ *diverges if* $|x| \geq |x_0|$.

To verify the condition, first notice that if $a_n x_0^n \nrightarrow 0$, then $a_n x^n \nrightarrow 0$ for all $|x| \geq |x_0|$, so the series surely diverges at x_0 and all $|x| \geq |x_0|$. Now suppose $a_n x_0^n \longrightarrow 0$. This does not imply that $\sum a_n x_0^n$ converges, but it does imply that $|a_n x_0^n| < 1$ for all large n. If $|x| < |x_0|$, so that $|\frac{x}{x_0}| < 1$, then for large n,

$$|a_n x^n| = |a_n x_0^n| \left|\frac{x}{x_0}\right|^n < \left|\frac{x}{x_0}\right|^n.$$

Since $\sum |\frac{x}{x_0}|^n$ is a convergent geometric series, $\sum |a_n x^n|$ converges if $|x| < |x_0|$; that is, $\sum a_n x^n$ converges absolutely if $|x| < |x_0|$.

We saw in Chapter 21 that $|n!x^n| \longrightarrow \infty$ for all $x \neq 0$. Hence, $\sum n!x^n$ converges only at $x = 0$. Such a series is of no use, since we want to use series to represent functions, and we will therefore now consider only series that converge for some nonzero values of x.

If a power series converges for some values of $x \neq 0$ and diverges for some x, then it follows from the Convergence Condition that there will either be a largest number r such that $a_n r^n \longrightarrow 0$, or a smallest number r such that $a_n r^n \not\rightarrow 0$. In either case, the series will converge absolutely if $|x| < r$ and diverge if $|x| > r$. The number r is called the **radius of convergence**, and the interval $(-r, r)$ is the **interval of convergence**. The series may converge at either or neither or both of the numbers r and $-r$. The convergence or divergence at these points is generally of no importance.

If $a_n x^n \longrightarrow 0$ for all x, then the series converges absolutely for all x, and we say the radius of convergence is ∞ and the interval of convergence is the whole line $(-\infty, \infty)$. For example, we saw in Chapter 21 that $\frac{x^n}{n!} \longrightarrow 0$ for all x, so $\sum \frac{x^n}{n!}$ converges absolutely for all x. We will show in Chapter 26 that

$$e^x = \sum_{n=0}^{\infty} \frac{x^n}{n!} = 1 + x + \frac{x^2}{2!} + \frac{x^3}{3!} + \frac{x^4}{4!} + \cdots .$$

Recall that the sequences $\{(\log n)^k\}$, $\{n^p\}$ with $p > 0$, $\{a^n\}$ with $a > 1$, and $\{n!\}$ represent different orders of growth in the sense that

$$\frac{(\log n)^k}{n^p} \longrightarrow 0, \quad \frac{n^p}{a^n} \longrightarrow 0, \quad \frac{a^n}{n!} \longrightarrow 0. \tag{24.2}$$

Since $\frac{n^p}{a^n} \longrightarrow 0$ for all $a > 1$, it follows, writing $x = \frac{1}{a}$, that $n^p x^n \longrightarrow 0$ for all $|x| < 1$.

EXAMPLE 24.1

Find the radius of convergence of $\sum \frac{n^2}{3^n} x^n = \sum n^2 (\frac{x}{3})^n$.

From (24.2) we know that $n^2 x^n \longrightarrow 0$ if $|x| < 1$, and of course $|n^2 x^n| \longrightarrow \infty$ if $|x| \geq 1$. Hence, $n^2 (\frac{x}{3})^n \longrightarrow 0$ if and only if $|\frac{x}{3}| < 1$. The radius of convergence is 3, and the series converges absolutely on $(-3, 3)$.

Coefficients that are powers of n, like n^2 in Example 24.1, have no effect on the radius of convergence. All three of the following series have the same radius, 3:

$$\sum n^2 \left(\frac{x}{3}\right)^n, \quad \sum \left(\frac{x}{3}\right)^n, \quad \sum \frac{1}{n^2} \left(\frac{x}{3}\right)^n .$$

This follows from the convergence condition because both $|n^2 (\frac{x}{3})^n|$ and $|\frac{1}{n^2} (\frac{x}{3})|$ approach 0 if $\left|\frac{x}{3}\right| < 1$, and both tend to ∞ if $|\frac{x}{3}| > 1$. A similar statement holds for $\log n$ or powers of $\log n$. For example, both $|(\log n)^k x^n|$ and $|\frac{1}{(\log n)^k} x^n|$ tend to zero if $|x| < 1$, and both tend to ∞ if $|x| > 1$. Therefore, all three of the following series have radius 1 for any value of k:

$$\sum (\log n)^k x^n, \quad \sum x^n, \quad \sum \frac{1}{(\log n)^k} x^n .$$

EXAMPLE 24.2

Find the radius of $\sum 2^n (\log n)^3 x^n$ and $\sum \frac{2^n}{(\log n)^3} x^n$.

We write the terms in the following form: $(\log n)^3 (2x)^n$ and $\frac{1}{(\log n)^3} (2x)^n$. Both terms approach zero for $|2x| < 1$, and both tend to ∞ if $|2x| > 1$. Therefore, the radius for both series is $\frac{1}{2}$.

The following test, the ratio test, is a popular technique because it requires very little thought. The ratio test applies to any series, so we state the test for a series of constants, $\sum a_n$, and then show how it applies to power series.

Ratio Test: *The series* $\sum a_n$ *converges absolutely if* $\lim_{n\to\infty} |\frac{a_{n+1}}{a_n}| < 1$ *and diverges if* $\lim_{n\to\infty} |\frac{a_{n+1}}{a_n}| > 1$.

To verify the ratio test, first notice that if $\lim_{n\to\infty}|\frac{a_{n+1}}{a_n}| > 1$, then $|a_{n+1}| > |a_n|$ for all large n, so $a_n \not\to 0$ and the series surely diverges. Now assume $\lim_{n\to\infty}|\frac{a_{n+1}}{a_n}| = r < 1$, and let s be a number between r and 1. Then $|\frac{a_{n+1}}{a_n}| < s$ for all sufficiently large n – say for all $n \geq N$. Hence,

$$\left|\frac{a_{N+1}}{a_N}\right| < s, \quad |a_{N+1}| < |a_N|s,$$

$$\left|\frac{a_{N+2}}{a_{N+1}}\right| < s, \quad |a_{N+2}| < |a_{N+1}|s < |a_N|s^2,$$

and in general $|a_{N+k}| < |a_N|s^k$. Since $\sum s^k$ and $\sum |a_N|s^k$ are convergent geometric series, $\sum |a_n|$ converges.

EXAMPLE 24.3

Use the ratio test to find the radius of convergence of $\sum \frac{\log n}{n^2 5^{n+2}} x^n$.

$$\lim_{n\to\infty}\left|\frac{\log(n+1)}{(n+1)^2 5^{n+3}}x^{n+1}\cdot\frac{n^2 5^{n+2}}{\log n}\frac{1}{x^n}\right|$$

$$= |x|\lim_{n\to\infty}\frac{n^2}{(n+1)^2}\frac{\log(n+1)}{\log n}\frac{1}{5} = \frac{|x|}{5}.$$

The series converges if $|\frac{x}{5}| < 1$; the radius is 5.

Example 24.3 could also be done simply by rewriting the terms:

$$\frac{\log n}{n^2 5^{n+2}}x^n = \frac{\log n}{25n^2}\left(\frac{x}{5}\right)^n.$$

The factor $\frac{\log n}{25n^2}$ does not change the radius from that of $\sum(\frac{x}{5})^n$. The ratio test provides a mechanical way of showing the insignificance of $\frac{\log n}{25n^2}$ relative to $(\frac{x}{5})^n$.

The next example shows a more realistic use of the ratio test.

EXAMPLE 24.4

Find the radius of convergence of $\sum \frac{n!}{n^n}x^n$.

We calculate the limiting ratio:

$$\lim_{n\to\infty}\left|\frac{(n+1)!}{(n+1)^{n+1}}x^{n+1}\cdot\frac{n^n}{n!}\frac{1}{x^n}\right|$$

$$= |x|\lim_{n\to\infty}\frac{(n+1)!}{n!}\cdot\frac{n^n}{(n+1)^{n+1}}$$

$$= |x|\lim_{n\to\infty}\frac{(n+1)\cdot n!}{n!}\cdot\frac{n^n}{(n+1)\cdot(n+1)^n}$$

$$= |x|\lim_{n\to\infty}\left(\frac{n}{n+1}\right)^n = |x|\lim_{n\to\infty}\frac{1}{(1+\frac{1}{n})^n} = \frac{|x|}{e}.$$

The radius is e.

From the preceding examples we see that for power series $\sum a_n x^n$, the ratio test always looks like this:

$$\lim_{n\to\infty}\left|\frac{a_{n+1}x^{n+1}}{a_n x^n}\right| = |x|\lim_{n\to\infty}\left|\frac{a_{n+1}}{a_n}\right|.$$

The radius is the reciprocal of $\lim_{n\to\infty}|\frac{a_{n+1}}{a_n}|$.

A power series represents a function on its interval of convergence, and much of the importance of power series derives from the fact that they can be differentiated and integrated term-by-term. That is, if

$$f(x) = \sum_{n=0}^{\infty} a_n x^n = a_0 + a_1 x + a_2 x^2 + a_3 x^3 + \cdots , \tag{24.3}$$

and the series converges on $(-r, r)$, then $f'(x)$ and $\int_0^x f(t)\,dt$ exist on $(-r, r)$, and

$$f'(x) = \sum_{n=1}^{\infty} n a_n x^{n-1} = a_1 + 2a_2 x + 3a_3 x^2 + \cdots , \tag{24.4}$$

$$\int_0^x f(t)dt = \sum_{n=0}^{\infty} \frac{a_n}{n+1} x^{n+1} = a_0 x + \frac{a_1}{2} x^2 + \frac{a_2}{3} x^3 + \cdots . \tag{24.5}$$

The radius does not change when you differentiate or integrate a series term-by-term, and the resulting series represent the right functions. If $f(x)$ has a power series representation, then $f(x)$ has derivatives of all orders, and all these derivatives have power series, and all the series have the same radius of convergence.

Now consider some of the functions we can represent with series, starting with the geometric series. If $|x| < 1$, then

$$\frac{1}{1-x} = \sum_{n=0}^{\infty} x^n = 1 + x + x^2 + x^3 + \cdots . \tag{24.6}$$

Since $|-x| < 1$ and $|\pm x^2| < 1$ if and only if $|x| < 1$, we can substitute in (24.6) to get

$$\frac{1}{1+x} = \sum_{n=0}^{\infty} (-x)^n = 1 - x + x^2 - x^3 + x^4 - \cdots , \tag{24.7}$$

$$\frac{1}{1-x^2} = \sum_{n=0}^{\infty} (x^2)^n = 1 + x^2 + x^4 + x^6 + \cdots , \tag{24.8}$$

$$\frac{1}{1+x^2} = \sum_{n=0}^{\infty} (-x^2)^n = 1 - x^2 + x^4 - x^6 + \cdots . \tag{24.9}$$

Now the series above can be differentiated or integrated, and for x in $(-1, 1)$ we get:

$$\log(1+x) = \int \frac{dx}{1+x} = x - \frac{1}{2}x^2 + \frac{1}{3}x^3 - \frac{1}{4}x^4 + \cdots , \tag{24.10}$$

$$\frac{1}{(1-x)^2} = \frac{d}{dx}\frac{1}{1-x} = 1 + 2x + 3x^2 + 4x^3 + \cdots , \tag{24.11}$$

$$\tan^{-1} x = \int \frac{dx}{1+x^2} = x - \frac{1}{3}x^3 + \frac{1}{5}x^5 - \frac{1}{7}x^7 + \cdots . \tag{24.12}$$

The series for $\log(1+x)$ and $\tan^{-1} x$ are proper alternating series when $x = 1$, and it can be shown that they converge to the right thing; that is,

$$\log 2 = 1 - \frac{1}{2} + \frac{1}{3} - \frac{1}{4} + \cdots , \tag{24.13}$$

$$\tan^{-1} 1 = \frac{\pi}{4} = 1 - \frac{1}{3} + \frac{1}{5} - \frac{1}{7} + \cdots . \tag{24.14}$$

If we substitute $3x$ for x in the series for $\frac{1}{1-x}$, we get

$$\frac{1}{1-3x} = 1 + (3x) + (3x)^2 + (3x)^3 + \cdots, \tag{24.15}$$

and this series converges provided $|3x| < 1$; that is, provided $|x| < \frac{1}{3}$. Similarly, if we replace x by $\frac{x}{2}$ in (24.11), we get

$$\frac{1}{\left(1-\frac{x}{2}\right)^2} = \frac{4}{(2-x)^2} = 1 + \frac{2x}{2} + \frac{3x^2}{4} + \frac{4x^3}{8} + \cdots,$$

and this series converges if $|\frac{x}{2}| < 1$, or $|x| < 2$.

EXAMPLE 24.5

Find a power series for $\log(1+3x)$ and find its radius of convergence.

From (24.10) we have, for $|x| < 1$,

$$\log(1+x) = x - \frac{1}{2}x^2 + \frac{1}{3}x^3 - \frac{1}{4}x^4 + \cdots.$$

We can substitute $3x$ for x as long as $|3x| < 1$, or $|x| < \frac{1}{3}$. Therefore,

$$\begin{aligned}\log(1+3x) &= 3x - \frac{1}{2}(3x)^2 + \frac{1}{3}(3x)^3 - \frac{1}{4}(3x)^4 + \cdots \\ &= 3x - \frac{3^2x^2}{2} + \frac{3^3x^3}{3} - \frac{3^4x^4}{4} + \cdots,\end{aligned}$$

and the radius of convergence is $\frac{1}{3}$.

EXAMPLE 24.6

We will show in Chapter 26 that

$$\sin x = x - \frac{1}{3!}x^3 + \frac{1}{5!}x^5 - \frac{1}{7!}x^7 + \cdots.$$

Use this to find a series for $\cos x$. What is the interval of convergence for both series?

Since $\frac{d}{dx}\sin x = \cos x$, we have

$$\begin{aligned}\cos x &= \frac{d}{dx}\left(x - \frac{x^3}{3!} + \frac{x^5}{5!} - \frac{x^7}{7!} + \cdots\right) \\ &= 1 - \frac{x^2}{2!} + \frac{x^4}{4!} - \frac{x^6}{6!} + \cdots. \qquad (24.16) \\ &= \sum_{n=0}^{\infty}(-1)^n\frac{x^{2n}}{(2n)!}.\end{aligned}$$

(0! is defined to be 1.) To find the interval of convergence, we use the ratio test:

$$\lim_{n\to\infty}\left|\frac{x^{2n+2}}{(2n+2)!}\cdot\frac{(2n)!}{x^{2n}}\right| = |x^2|\lim_{n\to\infty}\frac{1}{(2n+2)(2n+1)} = 0.$$

The limiting ratio is less than 1 for all x, so the series converges for all x.

EXAMPLE 24.7

Find a power series for $\cos\frac{x}{2}$.

Since the series for $\cos x$ converges for all x, we can substitute $\frac{x}{2}$ for x:

$$\cos\frac{x}{2} = 1 - \frac{1}{2!}\left(\frac{x}{2}\right)^2 + \frac{1}{4!}\left(\frac{x}{2}\right)^4 - \frac{1}{6!}\left(\frac{x}{2}\right)^6 + \cdots.$$

PROBLEMS

Find the radius of convergence.

24.1 $\sum n^3 x^n$

24.2 $\sum n^4 (2x)^n$

24.3 $\sum \frac{1}{n} \left(\frac{x}{4}\right)^n$

24.4 $\sum 2^n (\log n) x^n$

24.5 $\sum \frac{n^5}{5^n} x^n$

24.6 $\sum \frac{n + \sqrt{n}}{\log n} x^n$

24.7 $\sum \frac{n^3}{n!} x^n$

24.8 $\sum \frac{n^2 2^n}{n!} x^n$

24.9 $\sum \frac{n! x^n}{(2n)!}$

24.10 $\sum e^n x^n$

24.11 $\sum \frac{3^n}{n^2 + 1} \left(\frac{x}{2}\right)^n$

24.12 $\sum \frac{n! + n^2}{n! + \log n} x^n$

24.13 $\sum \frac{n! + 10n}{n! \cdot 10^n} x^n$

24.14 $\sum \frac{(\log n)^k}{n^p} x^n$ (k and p are arbitrary numbers.)

Find the interval of convergence.

24.15 $\sum n^3 (x - 3)^n$

24.16 $\sum \frac{2^n}{n} (x - 1)^n$

24.17 $\sum \frac{1}{n!} (x + 5)^n$

24.18 $\sum \frac{n^2}{5^n} (x + 1)^n$

Use the series given in the chapter for e^x, $\sin x$, $\cos x$, and the series derived from the series for $\frac{1}{1-x}$, to find the series for the given function by substitution, differentiation, or integration. Give the interval of convergence.

24.19 e^{2x}

24.20 e^{-x}

24.21 $\log(1 - x)$

24.22 $\log(1 + 2x)$

24.23 $\dfrac{1}{(1+x)^2}$

24.24 $\dfrac{1}{1+x^3}$

24.25 $\sin 2x$

24.26 $\tan^{-1}\left(\dfrac{x}{2}\right)$

24.27 $\cos x^2$

24.28 $\dfrac{x}{(1+x)^2}$

24.29 $\dfrac{\sin x}{x}$

24.30 $\dfrac{x}{1+x^2}$

24.31 $\log(1+x^2)$

24.32 Show that if $\lim_{n\to\infty} |\frac{a_{n+1}}{a_n}| = \frac{1}{r}$, so r is the radius of convergence of $\sum a_n x^n$, then the ratio test gives the same radius for $\sum n a_n x^{n-1}$ and $\sum \frac{a_n}{n+1} x^{n+1}$. *Hint*: For $x \neq 0$, $\sum n a_n x^{n-1}$ converges if and only if $\sum n a_n x^n$ converges, and $\sum \frac{a_n}{n+1} x^{n+1}$ converges if and only if $\sum \frac{a_n}{x+1}\, x^n$ converges.

25

Taylor Polynomials

In this chapter, we show how functions can usefully be approximated by polynomials. For example, there is no elementary antiderivative for e^{x^2}, so $\int_0^1 e^{x^2}\,dx$ can only be evaluated by some approximation method. If we can find (and we can) a polynomial $P(x)$ that is close to e^{x^2} on $[0, 1]$, then $\int_0^1 P(x)\,dx$ approximates $\int_0^1 e^{x^2}dx$. In the next chapter, we use these ideas to show how functions can be represented by power series.

Suppose we want to find an nth degree polynomial that approximates a function $f(x)$ as well as possible near zero. The reasonable thing to do is to find the polynomial that agrees with $f(x)$ at 0 and has the same derivatives as $f(x)$ at 0. All the derivatives of an nth degree polynomial are zero beyond the nth derivative, so we can only match the first n derivatives of $f(x)$ at 0 with those of an nth degree polynomial. If

$$P_n(x) = a_0 + a_1x + a_2x^2 + \cdots + a_nx^n,$$

then $P_n(x)$ has the following derivatives:

$$\begin{aligned}
P_n'(x) &= a_1 + 2a_2x + 3a_3x^2 + \cdots + na_nx^{n-1},\\
P_n''(x) &= 2a_2 + 3\cdot 2a_3x + 4\cdot 3a_4x^2 + \cdots + n(n-1)a_nx^{n-2},\\
P_n'''(x) &= 3\cdot 2a_3 + 4\cdot 3\cdot 2a_4x + 5\cdot 4\cdot 3a_5x^2 + \cdots + n(n-1)(n-2)a_nx^{n-3},\\
&\ \ \vdots\\
P_n^{(n)}(x) &= n!a_n.
\end{aligned}$$

Evaluating the derivatives at $x = 0$, we get

$$P_n(0) = a_0, \quad P_n'(0) = a_1, \quad P_n''(0) = 2a_2, \quad P_n'''(0) = 3\cdot 2a_3,$$

and in general

$$P_n^{(k)}(0) = k!a_k, \quad 0 \le k \le n.$$

To match up the derivatives of $P_n(x)$ with those of $f(x)$ at 0, we must have for each $k \le n$,

$$\begin{aligned}
P_n^{(k)}(0) &= k!a_k = f^{(k)}(0),\\
a_k &= \frac{1}{k!}f^{(k)}(0).
\end{aligned} \tag{25.1}$$

The equations (25.1) determine $a_0, a_1, a_2, \ldots, a_n$, so we know what $P_n(x)$ must be:

$$P_n(x) = f(0) + f'(0)x + \frac{1}{2!}f''(0)x^2 + \frac{1}{3!}f'''(0)x^3 + \cdots + \frac{1}{n!}f^{(n)}(0)x^n. \tag{25.2}$$

The polynomial (25.2), called the ***n*th Taylor polynomial for** $f(x)$ **at** 0, is the unique nth degree polynomial whose first n derivatives at zero match those of $f(x)$.

EXAMPLE 25.1

Find $P_n(x)$ for $f(x) = e^x$.

If $f(x) = e^x$, then $f^{(n)}(x) = e^x$ for all n, so $f^{(n)}(0) = 1$ for all n. Therefore,

$$P_n(x) = 1 + 1x + \frac{1}{2!}x^2 + \frac{1}{3!}x^3 + \cdots + \frac{1}{n!}x^n. \tag{25.3}$$

EXAMPLE 25.2

Find the general Taylor polynomial for $\sin x$ and for $\cos x$.

We calculate the successive derivatives of $\sin x$ at $x = 0$. This fortunately also gives us the derivatives of $\cos x$ at 0, since $\cos x$ is the first derivative of $\sin x$.

$$\begin{aligned}
f(x) &= \sin x, & f(0) &= 0, & a_0 &= 0,\\
f'(x) &= \cos x, & f'(0) &= 1, & a_1 &= 1,\\
f''(x) &= -\sin x, & f''(0) &= 0, & a_2 &= \frac{0}{2!},\\
f'''(x) &= -\cos x, & f'''(0) &= -1, & a_3 &= \frac{-1}{3!},\\
f^{(4)}(x) &= \sin x, & f^{(4)}(0) &= 0, & a_4 &= \frac{0}{4!},\\
f^{(5)}(x) &= \cos x, & f^{(5)}(0) &= 1, & a_5 &= \frac{1}{5!}.
\end{aligned} \tag{25.4}$$

The derivatives repeat after $n = 4$, and the pattern for the derivatives of $\sin x$ at 0 is $0, 1, 0, -1, 0, 1, 0, -1 \ldots$. Therefore, the Taylor polynomials for $\sin x$ are given by

$$P_{2n+1}(x) = x - \frac{x^3}{3!} + \frac{x^5}{5!} - \frac{x^7}{7!} + \cdots + \frac{(-1)^n x^{2n+1}}{(2n+1)!}. \tag{25.5}$$

The successive derivatives of $\cos x$ can also be read from the list (25.4); the values at zero are $1, 0, -1, 0, 1, 0, -1, 0, \ldots$. Therefore, the Taylor polynomials for $\cos x$ are

$$P_{2n}(x) = 1 - \frac{x^2}{2!} + \frac{x^4}{4!} - \frac{x^6}{6!} + \cdots + \frac{(-1)^n x^{2n}}{(2n)!}. \tag{25.6}$$

Now we will see what we can say about how well $P_n(x)$ agrees with $f(x)$ for values of x near zero. The graph of $P_1(x)$ is just the tangent line to $y = f(x)$ at $x = 0$. How close the linear function $P_1(x)$ is to $f(x)$ depends on how rapidly $f'(x)$ changes as x moves away from 0, and that depends on how large $f''(x)$ is. Suppose we have the estimate

$$m \le f''(x) \le M$$

on some interval $[0, a]$. If $R_1(x) = f(x) - P_1(x)$, then $R_1(0) = 0$ and $R_1'(0) = 0$ since $P_1(x)$ has the same value and slope as $f(x)$ at $x = 0$. Moreover, $R_1''(x) = f''(x)$, since $P_1''(x) = 0$.

Therefore, if $0 \le x \le a$,

$$\begin{aligned} m &\le R_1''(x) \le M, \\ \int_0^x m\,dx &\le \int_0^x R_1''(x)\,dx \le \int_0^x M\,dx, \\ mx &\le R_1'(x) \le Mx, \\ \int_0^x mx\,dx &\le \int_0^x R_1'(x)\,dx \le \int_0^x Mx\,dx, \\ \frac{1}{2}mx^2 &\le R_1(x) \le \frac{1}{2}Mx^2. \end{aligned} \tag{25.7}$$

Since $f(x) = P_1(x) + R_1(x)$, we can conclude from (25.7) that $f(x)$ lies between two parabolas over the interval $[0, a]$:

$$P_1(x) + \frac{1}{2}mx^2 \le f(x) \le P_1(x) + \frac{1}{2}Mx^2.$$

If we have an estimate on the size of $f'''(x)$—say $m \le f'''(x) \le M$ on $[0, a]$—then the argument above, with one more integration, shows that

$$\frac{1}{3!}mx^3 \le R_2(x) \le \frac{1}{3!}Mx^3,$$

and $f(x)$ lies between two cubics over $[0, a]$:

$$P_2(x) + \frac{m}{3!}x^3 \le f(x) \le P_2(x) + \frac{M}{3!}x^3. \tag{25.8}$$

In general, if $m \le f^{(n+1)}(x) \le M$ for $0 \le x \le a$, then on this interval

$$\frac{m}{(n+1)!}x^{n+1} \le R_n(x) \le \frac{M}{(n+1)!}x^{n+1}.$$

EXAMPLE 25.3

Use the fifth Taylor polynomial $P_5(x)$ for e^x to approximate $e^{\frac{1}{2}}$. Estimate the error.

From (25.3) we have

$$P_5(x) = 1 + x + \frac{1}{2!}x^2 + \frac{1}{3!}x^3 + \frac{1}{4!}x^4 + \frac{1}{5!}x^5,$$

from which we get the approximation

$$e^{\frac{1}{2}} \doteq P_5\left(\frac{1}{2}\right) = 1 + \frac{1}{2} + \frac{1}{2!}\left(\frac{1}{2}\right)^2 + \frac{1}{3!}\left(\frac{1}{2}\right)^3 + \frac{1}{4!}\left(\frac{1}{2}\right)^4 + \frac{1}{5!}\left(\frac{1}{2}\right)^5 = 1.64870.$$

Since $e < 3$, $e^{\frac{1}{2}} < 2$, and for $0 \le x \le \frac{1}{2}$, $1 \le \frac{d^6}{dx^6}e^x \le 2$. Therefore,

$$.00002 = \frac{1}{6!}\left(\frac{1}{2}\right)^6 < R_5\left(\frac{1}{2}\right) < \frac{2}{6!}\left(\frac{1}{2}\right)^6 = .00004,$$

and 1.64870 is accurate to four places. In fact, since the error is between .00002 and .00004, we can say $e^{\frac{1}{2}} = 1.64873 \pm .00001$.

EXAMPLE 25.4

Use $P_4(x)$ for e^x to approximate $\int_0^1 e^{x^2}\,dx$.

Since $1 \le e^x \le 3$ for $0 \le x \le 1$, we know that for these values of x,

$$P_4(x) + \frac{1}{5!}x^5 < e^x < P_4(x) + \frac{3}{5!}x^5. \tag{25.9}$$

Since $0 \le x^2 \le 1$ if $0 \le x \le 1$, we can substitute x^2 for x in (25.9), and get

$$P_4(x^2) + \frac{1}{5!}x^{10} < e^{x^2} < P_4(x^2) + \frac{3}{5!}x^{10},$$

$$\int_0^1 P_4(x^2)\,dx + \frac{1}{11 \cdot 5!} < \int_0^1 e^{x^2}\,dx < \int_0^1 P_4(x^2)\,dx + \frac{3}{11 \cdot 5!}.$$

Substituting x^2 for x in $P_4(x)$ and integrating, we get an approximation to the integral with error between $\frac{1}{11 \cdot 5!} = .00076$ and $\frac{3}{11 \cdot 5!} = .0023$. Rounding off, we would have to use .001 and .002 for lower and upper bounds on the error. Now calculate the integral of $P_4(x^2)$:

$$\int_0^1 P_4(x^2)\,dx = \int_0^1 \left(1 + x^2 + \frac{1}{2!}x^4 + \frac{1}{3!}x^6 + \frac{1}{4!}x^8\right) dx$$

$$= 1 + \frac{1}{3} + \frac{1}{5 \cdot 2!} + \frac{1}{7 \cdot 3!} + \frac{1}{9 \cdot 4!}$$

$$= 1.462.$$

Using the error estimates of .001 and .002, we can finally say that

$$1.463 < \int_0^1 e^{x^2}\,dx < 1.464.$$

If we wanted more accuracy, we could go to $P_5(x^2)$ and get an error between $\frac{1}{13 \cdot 6!} = .0001$ and $\frac{3}{13 \cdot 6!} = .0003$ (see Problem 25.10).

As a comparison, Simpson's Rule with $n = 10$, and carrying eight places for the arithmetic, gives 1.4626814 for $\int_0^1 e^{x^2}\,dx$. Of course, we have no way to estimate the accuracy of the Simpson's Rule number.

The preceding examples involving e^x are particularly simple since all the derivatives of e^x are the same, and we considered intervals to the right of 0. In case we are interested in negative x, or the derivatives of $f(x)$ have different signs, it is more convenient to use the following estimate for $R_n(x)$:

$$\textit{If } |f^{(n+1)}(x)| \le M \textit{ on } [-a, a], \textit{ then } |R_n(x)| \le \frac{M}{(n+1)!}|x|^{n+1}.$$

EXAMPLE 25.5

Calculate $\cos 1 = \cos 57.2958°$ to within .0005.

Let $f(x) = \cos x$, and notice that for all n, $f^{(n+1)}(x)$ equals $\pm \cos x$ or $\pm \sin x$. Therefore, $|f^{(n+1)}(x)| \le 1$ for all n and all x. Consequently, we have the following estimates:

$$|R_n(x)| \le \frac{1}{(n+1)!}|x|^{n+1}$$

$$|R_n(1)| \le \frac{1}{(n+1)!}.$$

We find the smallest n such that $\frac{1}{(n+1)!} < .0005$:

$$\frac{1}{3!} = .16, \quad \frac{1}{4!} = .04, \quad \frac{1}{5!} = .008, \quad \frac{1}{6!} = .0014, \quad \frac{1}{7!} = .0002.$$

Therefore, $P_6(1)$ will be within .0002 of cos 1. Notice, however, that $P_6(x) \equiv P_7(x)$ since the coefficient of x^7 is 0. Therefore, $R_7(1)$ will also give the difference between $P_6(1)$ and cos 1. $|R_7(1)| \le \frac{1}{8!} = .00002$, and

$$P_6(1) = P_7(1) = 1 - \frac{1}{2!} + \frac{1}{4!} - \frac{1}{6!} = .54028.$$

Therefore, $\cos 1 = .54028 \pm .00002$.

We can translate the whole approximation process to any x_0 and use polynomials in $(x - x_0)$ rather than polynomials in x. For example, it makes no sense to try and approximate $\log x$ with polynomials in x, since $\log 0$ is not defined. However, $\log x$ and its derivatives are easy to calculate at $x_0 = 1$, so it does make sense to use polynomials in $x - 1$. In general, the **nth Taylor polynomial for $f(x)$ at x_0** is

$$P_n(x) = f(x_0) + f'(x_0)(x - x_0) + \frac{1}{2!}f''(x_0)(x - x_0)^2 + \frac{1}{3!}f'''(x_0)(x - x_0)^3 + \cdots + \frac{1}{n!}f^{(n)}(x_0)(x - x_0)^n,$$

with the usual error estimate:
If $|f^{(n+1)}(x)| \leq M$ on $[x_0 - a, x_0 + a]$, then for x in this interval,

$$|R_n(x)| \leq \frac{M}{(n+1)!}|x - x_0|^{n+1}.$$

EXAMPLE 25.6
Calculate the nth Taylor polynomial at 1 for $\log x$ and find $\log 1.5$ to two decimal places.

Let $f(x) = \log x$, and calculate the derivatives at $x_0 = 1$:

$$\begin{aligned} f(x) &= \log x, & f(1) &= 0, \\ f'(x) &= x^{-1}, & f'(1) &= 1, \\ f''(x) &= -x^{-2}, & f''(1) &= -1, \\ f'''(x) &= 2x^{-3}, & f'''(1) &= 2, \\ f^{(4)}(x) &= -3!x^{-4}, & f^{(4)}(1) &= -3!, \end{aligned}$$

and in general we see that $f^{(n)}(1) = (-1)^{n+1}(n-1)!$. Hence,

$$\begin{aligned} P_n(x) &= (x-1) - \frac{1}{2!}(x-1)^2 + \frac{2}{3!}(x-1)^3 - \frac{3!}{4!}(x-1)^4 + \cdots + \frac{(-1)^{n+1}(n-1)!}{n!}(x-1)^n \\ &= (x-1) - \frac{1}{2}(x-1)^2 + \frac{1}{3}(x-1)^3 - \frac{1}{4}(x-1)^4 + \cdots + \frac{(-1)^{n+1}}{n}(x-1)^n. \end{aligned}$$

For $x \geq 1$, $|f^{(n+1)}(x)| = |n!x^{-(n+1)}| \leq n!$. Therefore, for $x \geq 1$, $|R_n(x)| \leq \frac{n!}{(n+1)!}(x-1)^{n+1}$. For $x = 1.5$,

$$|R_n(1.5)| \leq \frac{1}{n+1}(.5)^{n+1}.$$

We calculate values of $\frac{(.5)^{n+1}}{(n+1)}$ until we get a value less than .005:

$$\frac{(.5)^3}{3} = .04, \quad \frac{(.5)^4}{4} = .02, \quad \frac{(.5)^5}{5} = .006,$$

$$\frac{(.5)^6}{6} = .003.$$

Therefore, with an error less than .003,

$$\log 1.5 \doteq P_5(1.5) = \left(\frac{1}{2}\right) \cdot \frac{1}{2}\left(\frac{1}{2}\right)^2 + \frac{1}{3}\left(\frac{1}{2}\right)^3 - \frac{1}{4}\left(\frac{1}{2}\right)^4 + \frac{1}{5}\left(\frac{1}{2}\right)^5 = .407.$$

The calculator gives $\log 1.5 = .405$.

PROBLEMS

25.1 Find $P_1(x)$, $P_2(x)$, $P_3(x)$, and $P_4(x)$ for $f(x) = 5x^3 - 2x^2 + 3x - 1$.

25.2 Find $P_4(x)$ for $f(x) = \frac{1}{(1+x)}$.

25.3 Find $P_3(x)$ for $f(x) = \tan^{-1} x$.

25.4 (a) Find $P_4(x)$ for $f(x) = \sin x^2$.
(b) Substitute x^2 for x in $P_2(x)$ for $\sin x$.

25.5 Find $P_5(x)$ at $x_0 = 1$ for $f(x) = x^{\frac{1}{2}}$.

25.6 Let $P_4(x)$ be the fourth Taylor polynomial for e^x. Use $P_4(\frac{1}{3})$ to approximate $\sqrt[3]{e}$, and estimate the error.

25.7 Let $P_3(x)$ be the third Taylor polynomial for $\sin x$. Evaluate $\int_0^1 P_3(x^2)\,dx$ to approximate $\int_0^1 \sin x^2 dx$. Estimate the error. Notice that $P_3(x) = P_4(x)$.

25.8 Let $P_4(x)$ be the fourth Taylor polynomial for $\cos x$. Evaluate $\int_0^1 P_4(\sqrt{x})dx$ to approximate $\int_0^1 \cos\sqrt{x}\,dx$. Estimate the error.

25.9 Use Simpson's Rule with $n = 10$ to approximate $\int_0^1 e^{x^2} dx$. (See Example 25.4).

25.10 Let $P_5(x)$ be the fifth Taylor polynomial for e^x on [0, 1]. Calculate $\int_0^1 P_5(x^2)dx$ to estimate $\int_0^1 e^{x^2} dx$ and estimate the error.

26

Taylor Series

We have seen that if $f(x)$ has a power series representation,

$$f(x) = a_0 + a_1 x + a_2 x^2 + a_3 x^3 + \cdots + a_n x^n + \cdots,$$

then $f(x)$ has derivatives of all orders, and the derivatives have power series

$$f'(x) = a_1 + 2a_2 x + 3a_3 x^2 + 4a_4 x^3 + \cdots,$$
$$f''(x) = 2a_2 + 3 \cdot 2a_3 x + 4 \cdot 3a_4 x^2 + \cdots,$$

and all these power series have the same radius of convergence. By substituting 0 for x in the series for $f(x)$, $f'(x)$, $f''(x)$, and so on, we see that

$$a_0 = f(0), \quad a_1 = f'(0), \quad a_2 = \frac{1}{2!} f''(0), \ldots$$

and in general

$$a_n = \frac{1}{n!} f^{(n)}(0). \tag{26.1}$$

Formula (26.1) shows that the a_n are determined by the function, so a power series representation is unique.

If $f(x)$ is any function with derivatives of all orders at zero, we can form the power series with the coefficients (26.1), and the result is the **Taylor series for $f(x)$ in powers of x**:

$$\sum_{n=0}^{\infty} \frac{1}{n!} f^{(n)}(0) x^n. \tag{26.2}$$

This series does not always converge to $f(x)$. For example, if $f(x) = e^{-\frac{1}{x^2}}$ for $x \neq 0$, and $f(0) = 0$, then $f^{(n)}(0) = 0$ for all n. The Taylor series for this function is identically zero, but the function is obviously positive if $x \neq 0$. This function is severely atypical—all the common elementary functions of calculus do have a Taylor series that converges to the function on *some* interval.

We will show that the Taylor series (26.2) converges to $f(x)$ for several elementary functions. Let $P_n(x)$ be the nth partial sum of the series (26.2):

$$P_n(x) = f(0) + f'(0)x + \frac{1}{2!}f''(0)x^2 + \cdots + \frac{1}{n!}f^{(n)}(0)x^n. \tag{26.3}$$

Notice that $P_n(x)$ is the same as the nth Taylor polynomial for $f(x)$. We show that $P_n(x) \longrightarrow f(x)$ by showing that $R_n(x) = f(x) - P_n(x) \longrightarrow 0$. Recall that if $|f^{(n+1)}(x)| \leq M$ on an interval $[-a, a]$, then for these values of x,

$$|R_n(x)| \leq \frac{M}{(n+1)!}|x^{n+1}| \leq \frac{M}{(n+1)!}a^{n+1}. \tag{26.4}$$

If all of the derivatives $f^{(n)}(x)$ satisfy the same estimate, $|f^{(n)}(x)| \leq M$ on $[-a, a]$—that is, if the bound M does not depend on n—then $R_n(x) \longrightarrow 0$ for all $x \in [-a, a]$, since $\frac{a^n}{n!} \longrightarrow 0$ for any a.

Consider the functions $\sin x$ and $\cos x$. If $f(x) = \sin x$ or $f(x) = \cos x$, then $|f^{(n)}(x)| \leq 1$ for all x and all n, so for both $\sin x$ and $\cos x$,

$$|R_n(x)| \leq \frac{|x|^{n+1}}{(n+1)!}, \tag{26.5}$$

and so $P_n(x) \longrightarrow f(x)$ for all x for both $f(x) = \sin x$ and $f(x) = \cos x$. Using the Taylor polynomials from the last chapter, we see that $\sin x$ has the Taylor series

$$\sin x = x - \frac{x^3}{3!} + \frac{x^5}{5!} - \frac{x^7}{7!} + \cdots, \tag{26.6}$$

and the series converges for all x. Similarly, the Taylor polynomials for $\cos x$ converge to $\cos x$ for all x, and $\cos x$ has the Taylor series

$$\cos x = 1 - \frac{x^2}{2!} + \frac{x^4}{4!} - \frac{x^6}{6!} + \cdots. \tag{26.7}$$

EXAMPLE 26.1

Find $\cos(\frac{1}{2})$ to within .00005.

We could use the remainder estimate, but we get the same estimate very simply now that we know the series (26.7) converges to $\cos x$. The series (26.7) is a proper alternating series for $|x| < 1$, so the error is less than the first term omitted. We calculate several terms at $x = \frac{1}{2}$:

$$\frac{1}{2!}\left(\frac{1}{2}\right)^2 \doteq .13; \quad \frac{1}{4!}\left(\frac{1}{2}\right)^4 \doteq .003; \quad \frac{1}{6!}\left(\frac{1}{2}\right)^6 \doteq .00002.$$

Hence,

$$\cos\frac{1}{2} \doteq 1 - \frac{1}{2!}\left(\frac{1}{2}\right)^2 + \frac{1}{4!}\left(\frac{1}{2}\right)^4 \doteq .87760.$$

The error will be negative, since the first term omitted is negative, so

$$.87760 > \cos\frac{1}{2} > .87758.$$

The Taylor polynomials for e^x are

$$P_n(x) = 1 + x + \frac{x^2}{2!} + \frac{x^3}{3!} + \cdots + \frac{x^n}{n!},$$

and

$$|R_n(x)| \leq \frac{e^x|x|^{n+1}}{(n+1)!}. \tag{26.8}$$

On any fixed interval $[-a, a]$, $|e^x| \le e^a$, $|x|^{n+1} \le a^{n+1}$, so on $[-a, a]$,

$$|R_n(x)| \le \frac{e^a a^{n+1}}{(n+1)!} \longrightarrow 0.$$

Thus, $P_n(x) \longrightarrow e^x$ on every interval $[-a, a]$, and the Taylor series for e^x converges to e^x for all x:

$$e^x = 1 + x + \frac{x^2}{2!} + \frac{x^3}{3!} + \cdots + \frac{x^n}{n!} + \cdots. \tag{26.9}$$

EXAMPLE 26.2

Find the Taylor series for e^{x^2}.

Since the series (26.9) converges for all x, we can substitute anything for x in (26.9). In particular, substituting x^2 for x, we get

$$e^{x^2} = 1 + x^2 + \frac{x^4}{2!} + \frac{x^6}{3!} + \frac{x^8}{4!} + \cdots. \tag{26.10}$$

Since (26.10) is a power series and power series representations are unique, (26.10) is *the* power series for e^{x^2}.

EXAMPLE 26.3

Find the Taylor series for $\sqrt{1+x}$ and find the radius of convergence.

We calculate the derivatives at 0 and the coefficients a_n:

$$f(x) = (1+x)^{\frac{1}{2}}; \qquad f(0) = 1; \qquad a_0 = 1$$

$$f'(x) = \frac{1}{2}(1+x)^{-\frac{1}{2}}; \qquad f'(0) = \frac{1}{2}; \qquad a_1 = \frac{1}{2}$$

$$f''(x) = \frac{1}{2}\left(-\frac{1}{2}\right)(1+x)^{-\frac{3}{2}}; \qquad f''(0) = \frac{1}{2}\left(-\frac{1}{2}\right); \qquad a_2 = \frac{1}{2!}\left(\frac{1}{2}\right)\left(-\frac{1}{2}\right)$$

$$f'''(x) = \frac{1}{2}\left(-\frac{1}{2}\right)\left(-\frac{3}{2}\right)(1+x)^{-\frac{5}{2}}; \qquad f'''(0) = \frac{1\cdot 3}{2^3}; \qquad a_3 = \frac{1}{3!}\,\frac{1\cdot 3}{2^3}$$

$$f^{(4)}(x) = -\frac{1\cdot 3\cdot 5}{2^4}(1+x)^{-\frac{7}{2}}; \qquad f^{(4)}(0) = -\frac{1\cdot 3\cdot 5}{2^4}; \qquad a_4 = \frac{-1}{4!}\cdot\frac{1\cdot 3\cdot 5}{2^4}.$$

In general, we see that

$$a_n = (-1)^{n+1}\frac{1\cdot 3\cdot 5\cdot\ \cdots\ (2n-3)}{n!\cdot 2^n}, \tag{26.11}$$

and

$$f^{(n+1)}(x) = \pm\frac{1\cdot 3\cdot 5\cdot\ \cdots\ \cdot(2n-1)}{2^{n+1}}\cdot\frac{1}{(1+x)^{\frac{(2n+1)}{2}}}. \tag{26.12}$$

If $0 \le x < 1$, then $(1+x)^{\frac{-(2n+1)}{2}} \le 1$, and

$$|R_n(x)| \le \frac{1\cdot 3\cdot 5\cdot\ \cdots\ \cdot(2n-1)}{2^{n+1}}\,\frac{|x|^{n+1}}{(n+1)!}. \tag{26.13}$$

For $n = 2, 3, 4, \ldots$, the coefficients of $|x|^{n+1}$ in (26.13) are

$$\frac{1}{2^2}\cdot\frac{1}{2} = .13, \qquad \frac{1\cdot 3}{2^3}\cdot\frac{1}{6} = .06; \qquad \frac{1\cdot 3\cdot 5}{2^4}\cdot\frac{1}{24} = .04;$$

and so on. The coefficient of $|x|^{n+2}$ is obtained by multiplying the coefficient of $|x|^{n+1}$ by

$$\frac{2n+1}{2\cdot(n+2)} = \frac{2n+1}{2n+4} < 1,$$

so $|R_n(x)| \le a_n|x|^{n+1}$ where $a_n \le a_2 = .13$ for all $n \ge 2$. Clearly, $R_n(x) \longrightarrow 0$ if $0 \le x < 1$, and the Taylor series for $\sqrt{1+x}$ converges to the function on $[0, 1)$. The series therefore converges on $(-1, 1)$, since power series converge on intervals symmetric about the origin. The series does converge to $\sqrt{1+x}$ also on $(-1, 0]$, although the argument above does not include this case. Hence, for $-1 < x < 1$,

$$(1+x)^{\frac{1}{2}} = 1 + \frac{1}{2}x - \frac{1}{2!}\frac{1}{2^2}x^2 + \frac{1}{3!}\frac{1\cdot 3}{2^3}x^3 - \frac{1}{4!}\frac{1\cdot 3\cdot 5}{2^4}x^4 + \cdots . \tag{26.14}$$

Recall the binomial expansion for a positive integer exponent n:

$$\begin{aligned}(1+x)^n &= 1 + nx + \frac{n(n-1)}{2!}x^2 + \frac{n(n-1)(n-2)}{3!}x^3 \\ &\quad + \frac{n(n-1)(n-2)(n-3)}{4!}x^4 + \cdots + x^n.\end{aligned} \tag{26.15}$$

If we formally substitute $\frac{1}{2}$ for n in (26.15), we get the Taylor series (26.14) for $(1+x)^{\frac{1}{2}}$. If n is a positive integer, the binomial expansion (26.15) is an nth degree polynomial. If n is a fraction, the binomial expansion is an infinite series.

The Taylor series for $f(x)$ in powers of $(x - x_0)$ is $\sum_{n=0}^{\infty} \frac{1}{n!} f^{(n)}(x_0)(x - x_0)^n$. This is the series whose partial sums are the Taylor polynomials at x_0.

EXAMPLE 26.4

Find the Taylor series for e^x in powers of $x - 2$.

If $f(x) = e^x$, then $f^{(n)}(x) = e^x$ for all n, and so $f^{(n)}(2) = e^2$ for all n. Therefore, $a_n = \frac{e^2}{n!}$ and

$$e^x = 1 + e^2(x-2) + \frac{e^2}{2!}(x-2)^2 + \frac{e^2}{3!}(x-2)^3 + \cdots . \tag{26.16}$$

We could also obtain the result (26.16) as follows. Substitute $(x-2)$ for x in the series for e^x and get

$$e^{x-2} = 1 + (x-2) + \frac{1}{2!}(x-2)^2 + \frac{1}{3!}(x-2)^3 + \cdots .$$

Since $e^{x-2} = \frac{e^x}{e^2}$, $e^x = e^2 \cdot e^{x-2}$, and

$$e^x = e^2\left[1 + (x-2) + \frac{1}{2!}(x-2)^2 + \frac{1}{3!}(x-2)^3 + \cdots\right]$$

which is the same as (26.16).

EXAMPLE 26.5

Find the Taylor series for $\frac{1}{x}$ in powers of $(x-1)$.

We compute the coefficients as follows:

$$\begin{array}{lll} f(x) = x^{-1}; & f(1) = 1; & a_0 = 1; \\ f'(x) = -x^{-2}; & f'(1) = -1; & a_1 = -1; \\ f''(x) = 2x^{-3}; & f''(1) = 2; & a_2 = \dfrac{2}{2!} = 1; \\ f'''(x) = -3!x^{-4}, & f'''(1) = -3!; & a_3 = -1. \end{array}$$

It is easy to see the general pattern:

$$f^{(n)}(x) = (-1)^n n! x^{-(n+1)}, \quad a_n = (-1)^n,$$

so the Taylor series is

$$\frac{1}{x} = 1 - (x-1) + (x-1)^2 - (x-1)^3 + \cdots . \tag{26.17}$$

Notice that we can also easily obtain (26.17) from the geometric series for $\frac{1}{1+x}$:

$$\frac{1}{1+x} = 1 - x + x^2 - x^3 + x^4 - \cdots ,$$

so if $|x-1|<1$,

$$\frac{1}{x}=\frac{1}{1+(x-1)}=1-(x-1)+(x-1)^2-(x-1)^3+\cdots.$$

EXAMPLE 26.6

Find the Taylor series at 1 for $\log x$.

We simply integrate the series (26.17):

$$\log x=\int_1^x \frac{1}{t}\,dt=(x-1)-\frac{1}{2}(x-1)^2+\frac{1}{3}(x-1)^3-\cdots. \tag{26.18}$$

EXAMPLE 26.7

Use the series for $\sin x$ to show that $\frac{\sin x}{x}\longrightarrow 1$ and $\frac{\sin x-x}{x^3}\longrightarrow -\frac{1}{6}$ as $x\longrightarrow 0$.

Since

$$\sin x=x-\frac{x^3}{3!}+\frac{x^5}{5!}-\frac{x^7}{7!}+\cdots,$$

we have, for all $x\neq 0$,

$$\frac{\sin x}{x}=1-\frac{x^2}{3!}+\frac{x^4}{5!}-\frac{x^6}{7!}+\cdots,$$

so $\frac{\sin x}{x}\longrightarrow 1$ as $x\longrightarrow 0$. Also

$$\begin{aligned}\frac{\sin x-x}{x^3}&=\frac{1}{x^3}\left[-\frac{x^3}{3!}+\frac{x^5}{5!}-\frac{x^7}{7!}+\cdots\right]\\&=-\frac{1}{3!}+\frac{x^2}{5!}-\frac{x^4}{7!}+\cdots,\end{aligned}$$

and hence, as $x\longrightarrow 0$,

$$\frac{\sin x-x}{x^3}\longrightarrow -\frac{1}{3!}.$$

PROBLEMS

26.1 Find $\cos\frac{1}{4}$ to within .0005. Do you need more terms for this accuracy or fewer than for $\cos\frac{1}{2}$ as in Example 26.1?

26.2 Find $\sin\frac{1}{10}$ to within 10^{-7}.

26.3 Evaluate $e^{-\frac{1}{4}}$ to within .0002.

26.4 Find five terms (through x^4) of the Taylor series for $(1+x)^{\frac{1}{3}}$, and check with the binomial expansion.

26.5 Use the Taylor series (26.14) for $\sqrt{1+x}$ to obtain the Taylor series for $\sqrt{1-x}$ and for $\sqrt{1+x^2}$.

26.6 Find the Taylor series for e^x in powers of $x-3$ and in powers of $x+1$.

26.7 Find the Taylor series for $\frac{1}{x}$ in powers of $(x-2)$ two ways:
(i) Calculate $\frac{1}{n!}f^{(n)}(2)$ for all n;
(ii) use $\frac{1}{x}=\frac{1}{2}\frac{1}{1+(\frac{x-2}{2})}$.

26.8 Find the Taylor series for $\log x$ in powers of $x-2$. *Hint*: $\log x=\log 2+\int_2^x\frac{dt}{t}$ so you can integrate the series in Problem 26.7.

26.9 Use the series for $\cos x$ to show that $\frac{(\cos x-1)}{x^2}\longrightarrow -\frac{1}{2}$ as $x\longrightarrow 0$.

26.10 For $x\neq 0$,

$$\frac{\sin x}{x}=1-\frac{x^2}{3!}+\frac{x^4}{5!}-\frac{x^6}{7!}+\cdots.$$

Therefore, for $0 < x \leq 1$,

$$\left|\frac{\sin x}{x} - \left[1 - \frac{x^2}{3!} + \frac{x^4}{5!}\right]\right| \leq \frac{x^6}{7!}.$$

Use this to approximate $\int_0^1 \frac{\sin x}{x}\,dx$ to within $\int_0^1 \frac{x^6}{7!}\,dx \doteq .00003$. *Hint*: Since $\frac{\sin x}{x} \longrightarrow 1$ as $x \longrightarrow 0+$, you can regard this as a proper integral.

27

Separable Differential Equations

Many physical facts attain their mathematical expression as differential equations. In the simplest case, a **differential equation** is an equation expressing a relationship between two variables and the rate of change (derivative) of one variable with respect to the other. For example, $\frac{dy}{dx} = 3x^2$ is a simple differential equation, and the solutions are all the functions $y = x^3 + c$. All the indefinite integration problems we have treated can be considered differential equations in this way. Thus, "find $\int f(x)\,dx$" means the same as "solve the differential equation $\frac{dy}{dx} = f(x)$." The solutions are all the functions $y = \int f(x)\,dx + c$. Notice that now we want *all* solutions, so the arbitrary constant is necessary. In general, the solutions of a differential equation represent a *family* of curves, and not just a single curve. To specify a specific solution, we must specify an **initial condition** like $y(x_0) = y_0$.

EXAMPLE 27.1

Find all solutions of the differential equation, and find the specific solution that satisfies the given initial condition: $\frac{dy}{dx} = \frac{1}{4+x^2}$, $y(0) = 1$.

Solution

$$\frac{dy}{dx} = \frac{1}{4+x^2},$$

$$y = \int \frac{1}{4+x^2}\,dx = \frac{1}{2}\tan^{-1}\frac{x}{2} + c.$$

This is the family of all solutions. To find c so that the initial condition is satisfied, substitute 0 for x and 1 for y:

$$1 = \frac{1}{2}\tan^{-1}\frac{0}{2} + c; \quad c = 1.$$

The solution that satisifies the initial condition is

$$y = \frac{1}{2}\tan^{-1}\frac{x}{2} + 1.$$

A **first-order** differential equation is one that involves only the first derivative. A first-order differential equation generally has a one-parameter family of solutions; that is, the solutions depend on a single arbitrary constant. In this chapter, we will study the following useful type of first-order equation:

$$\frac{dy}{dx} = \frac{f(x)}{g(y)},$$

or equivalently,

$$g(y)\frac{dy}{dx} = f(x). \tag{27.1}$$

An equation of this type is called **separable**, and we say the **variables separate** because we can write all the x's on one side and all the y's on the other. We bend the notation a little and agree that (27.1) can also be written

$$g(y)\,dy = f(x)\,dx. \tag{27.2}$$

As this last form suggests, the solutions of (27.1) or (27.2) are obtained simply by integrating:

$$\int g(y)\,dy = \int f(x)\,dx + c. \tag{27.3}$$

That is, the solutions y of (27.2) will be the functions that are defined implicitly by one of the equations

$$G(y) = F(x) + c$$

where G and F are antiderivatives of g and f, respectively.

EXAMPLE 27.2

Solve $\dfrac{dy}{dx} = \dfrac{y^2}{x+1}$.

Solution

We separate the variables and integrate:

$$\frac{1}{y^2}\,dy = \frac{1}{x+1}\,dx,$$

$$-\frac{1}{y} = \log|x+1| + c.$$

Since c is arbitrary, we can write $\log c$ instead of c to simplify the expression:

$$-\frac{1}{y} = \log|x+1| + \log c = \log c|x+1|,$$

$$y = \frac{-1}{\log c|x+1|}.$$

The process of separating the variables and integrating appears to be a purely formal one, but it really does give all the solutions. It is easy to see by differentiating both sides that if y satisfies an equation

$$G(y) = F(x) + c, \tag{27.4}$$

where $G'(y) = g(y)$ and $F'(x) = f(x)$, then y also satisfies the differential equation

$$g(y)\frac{dy}{dx} = f(x). \tag{27.5}$$

It is also easy to show (Problem 27.19) that there are no solutions of the differential equation (27.5) other than functions that satisfy (27.4).

One very common separable differential equation is the exponential change equation

$$\frac{dy}{dx} = ay, \tag{27.6}$$

where a is a given constant. The equation expresses the fact that y increases ($a > 0$) or decreases ($a < 0$) at a rate proportional to y. This equation characterizes a number of physical phenomena, ranging from the growth of bacteria colonies to the decay of radioactive substances. We can separate variables and integrate to solve (27.6), but it is simpler just to notice that the functions $y = ce^{ax}$ all satisfy the differential equation. Moreover, if y is any solution, (i.e., any function such that $\frac{dy}{dx} = ay$), then

$$\begin{aligned}\frac{d}{dx}(ye^{-ax}) &= \frac{dy}{dx} \cdot e^{-ax} - aye^{-ax} \\ &= ay \cdot e^{-ax} - aye^{-ax} = 0.\end{aligned}$$

Since $\frac{d}{dx}(ye^{-ax}) = 0$, ye^{-ax} is constant, and

$$y = ce^{ax}. \tag{27.7}$$

That is, every function (27.7) satisfies the differential equation, and any function that satisfies the differential equation has the form (27.7).

EXAMPLE 27.3

Let y be the number of bacteria in a colony at time t, and assume $\frac{dy}{dt} = ay$, so $y = ce^{at}$ for some constants c and a. If $y = 500$ at $t = 0$ and $y = 2000$ at $t = 2$ hours, what is y at $t = 4$ hours?

Solution

Since $y = 500$ at $t = 0$, we have $500 = ce^0 = c$ and $c = 500$. Notice that c will always be the value of y at $t = 0$ in exponential change problems. Now we use the fact that $y = 2000$ when $t = 2$ to find a, or more usefully, to find e^a:

$$\begin{aligned}2000 &= 500e^{2a} = 500(e^a)^2, \\ (e^a)^2 &= 4, \\ e^a &= 2, \quad e^{at} = 2^t.\end{aligned}$$

The final solution is $y = 500 \cdot 2^t$. When $t = 4$, $y = 500 \cdot 2^4 = 8000$.

EXAMPLE 27.4

Solve $\dfrac{dy}{dx} = 2xy + 2x$.

Solution

$$\begin{aligned}\frac{dy}{dx} &= 2x(y+1), \\ \frac{dy}{y+1} &= 2x\,dx, \\ \log|y+1| &= x^2 + \log c.\end{aligned}$$

We write the arbitrary constant in the form $\log c$ and use $e^{\log c} = c$ to simplify the expression:

$$\begin{aligned}|y+1| &= e^{x^2} \cdot e^{\log c} = ce^{x^2}, \\ y + 1 &= \pm ce^{x^2}.\end{aligned}$$

We now allow c to be positive or negative and we drop the $\pm$ sign, so finally

$$y = -1 + ce^{x^2}.$$

EXAMPLE 27.5

Newton's law of cooling says that a hot body will cool off at a rate proportional to the difference between its temperature and the temperature of the surroundings. Suppose a cup of coffee has temperature $T = 130°$F at $t = 0$ in a room at temperature 70°F. If $T = 100°$F at $t = 3$ minutes, what is the temperature at $t = 5$ minutes? at $t = 6$ minutes?

Solution

The cooling law gives the differential equation

$$\frac{dT}{dt} = -k(T - 70),$$

where T is the temperature in degrees Fahrenheit and k is a positive constant of proportionality.

We separate the variables and solve.

$$\frac{dT}{T - 70} = -k\,dt,$$
$$\log(T - 70) = -kt + \log c,$$
$$T - 70 = ce^{-kt}.$$

Since $T = 130$ at $t = 0$, $c = 60$ and

$$T = 70 + 60e^{-kt}.$$

Now we use the condition $T = 100$ when $t = 3$ to find e^{-k}.

$$100 = 70 + 60e^{-3k},$$
$$\frac{30}{60} = (e^{-k})^3,$$
$$e^{-k} = 2^{-\frac{1}{3}}, \quad e^{-kt} = 2^{-t/3}.$$

Finally, we have

$$T = 70 + 60 \cdot 2^{-t/3}.$$

When $t = 5$ minutes, this gives

$$T = 70 + \frac{60}{2^{\frac{5}{3}}} \doteq 70 + 19 = 89.$$

When $t = 6$ minutes,

$$T = 70 + \frac{60}{2^{\frac{6}{3}}} = 70 + 15 = 85.$$

A body falling in a vacuum accelerates at a constant rate of 32 ft/sec^2. Since tall vacuums are rare, air resistance usually plays a significant role, with the acceleration usually being retarded at a rate proportional to the speed. Thus, a more realistic differential equation for the velocity v of a falling object is

$$\frac{dv}{dt} = 32 - kv, \tag{27.8}$$

where the constant k depends on the shape of the object. In Problem 27.18 you are asked to solve (27.8) and show that

$$v = \frac{32}{k}(1 - e^{-kt}),$$

from which we conclude that v approaches a **terminal velocity** of $\frac{32}{k}$ ft/sec.

PROBLEMS

Solve the differential equation. Find the constant if an initial condition is given.

27.1 $\dfrac{dy}{dx} = x\cos x^2;\quad y(0) = 2$

27.2 $\dfrac{dy}{dx} = \sec^2 x;\quad y\left(\dfrac{\pi}{4}\right) = 0$

27.3 $\frac{dy}{dx} = \frac{3}{x^2+9}$; $y(3) = \frac{\pi}{2}$

27.4 $\frac{dy}{dx} = \frac{x}{x^2+1}$; $y(0) = 5$

27.5 $\frac{dy}{dx} = \frac{x^2+1}{y}$; $y(0) = 2$

27.6 $\frac{dy}{dx} = e^{x-y}$; $y(0) = \log 2$

27.7 $\frac{dy}{dx} = \frac{4+y^2}{1+x^2}$

27.8 $\frac{dy}{dx} = \frac{y}{\sqrt{1-x^2}}$

27.9 $\frac{dy}{dx} - 5y = 0$

27.10 $\frac{dy}{dx} - 5y = 10$

27.11 $\frac{dy}{dx} - 2xy = 0$

27.12 $\frac{dy}{dx} - 2xy = 2x$

27.13 $\frac{dy}{dx} = \frac{y}{x} + y$

27.14 $\frac{dy}{dx} = \frac{y+2}{x-1}$

27.15 Suppose a beaker of boiling water at 100°C cools to 40°C in five minutes in a room at 20°C. What is its temperature T at time t? at time $t = 10$ minutes?

27.16 Suppose there are 100 bacteria in a colony at $t = 0$ and 500 three hours later. When will there be 1000 bacteria in the colony? When will there be 2500?

27.17 Show that the solutions of $\frac{dy}{dt} = ay$ can be written $y = y_0 A^t$ where y_0 is the initial value and A is a constant.

27.18 Solve equation (27.8) for a falling body with air resistance proportional to speed, and $v = 0$ when $t = 0$, and show that $v \longrightarrow \frac{32}{k}$ as $t \longrightarrow \infty$. (A skydiver with unopened chute reaches a terminal velocity of about 120 mph, or about 176 ft/sec, which gives $k = .18$.)

27.19 Show that if y satisfies $g(y)\frac{dy}{dx} = f(x)$, then y must satisfy an equation $G(y) = F(x) + c$, with $G' = g$ and $F' = f$. *Hint*: Define $H(x) = G(y(x))$ and show that $H'(x) = F'(x)$, so $H(x) = F(x) + c$.

28

First-Order Linear Equations

A **first-order linear** differential equation is one that can be written in the following form:

$$\frac{dy}{dx} + p(x)y = q(x), \tag{28.1}$$

where $p(x)$ and $q(x)$ are given functions. The equation is called **linear** because if we let $L(y)$ denote the left side of (28.1), then for any constant c and any two functions y_1 and y_2 we have:

$$\begin{aligned} L(cy_1) &= cL(y_1), \\ L(y_1 + y_2) &= L(y_1) + L(y_2). \end{aligned} \tag{28.2}$$

If $q(x) \equiv 0$ the equation (28.1) is called **homogeneous**, and the equation looks like this:

$$L(y) = \frac{dy}{dx} + p(x)y = 0. \tag{28.3}$$

The homogeneous equation (28.3) is called the **reduced form** of equation (28.1). Notice that if y_0 is a solution of (28.1), so that $L(y_0) = q(x)$, and y_1 is a solution of the reduced equation (28.3), so that $L(y_1) = 0$, then $y_0 + cy_1$ is again a solution of (28.1), since

$$L(y_0 + cy_1) = L(y_0) + cL(y_1) = q(x) + c \cdot 0.$$

The process of solving the linear equation (28.1) consists of finding one solution y_0 and adding to it all solutions cy_1 of the reduced equation.

The following observation is the key to finding the solutions of (28.1) and (28.3). Let $P(x) = \int p(x)\,dx$, so that $P'(x) = p(x)$, and check that

$$\begin{aligned} \frac{d}{dx}\left[e^{P(x)}y\right] &= e^{P(x)}\frac{dy}{dx} + e^{P(x)}p(x)y \\ &= e^{P(x)}\left[\frac{dy}{dx} + p(x)y\right]. \end{aligned} \tag{28.4}$$

This calculation shows that the left side of (28.1) becomes exactly the derivative of $e^{P(x)}y$ if we multiply by $e^{P(x)}$. Multiplying *both* sides of (28.1) by $e^{P(x)}$, we get the equivalent equation

$$e^{P(x)}\left[\frac{dy}{dx} + p(x)y\right] = e^{P(x)}q(x),$$

which because of (28.4) can be written

$$\frac{d}{dx}\left[e^{P(x)}y\right] = e^{P(x)}q(x). \tag{28.5}$$

The solutions of (28.5), and hence the solutions of (28.1), since (28.1) and (28.5) are equivalent, can now be obtained by integrating both sides of (28.5):

$$\begin{aligned} e^{P(x)}y &= \int e^{P(x)}q(x)\,dx + C, \\ y &= e^{-P(x)}\int e^{P(x)}q(x)\,dx + Ce^{-P(x)}. \end{aligned} \tag{28.6}$$

If $q(x) \equiv 0$, then (28.6) gives the general solution $y = Ce^{-P(x)}$ of the reduced equation. The expression

$$y_0 = e^{-P(x)}\int e^{P(x)}q(x)\,dx$$

is the one particular solution of (28.1) that we need to go with the general solution $ce^{-P(x)}$ of the reduced equation.

Rather than memorizing the formula (28.6), it is much easier to multiply both sides of the equation by $e^{P(x)}$, and integrate.

EXAMPLE 28.1

Solve $\dfrac{dy}{dx} + 3x^2y = 0.$

Solution

Here $p(x) = 3x^2$, $P(x) = x^3$, and we multiply both sides by e^{x^3}:

$$\begin{aligned} e^{x^3}\frac{dy}{dx} + e^{x^3}\cdot 3x^2y &= 0, \\ \frac{d}{dx}\left[e^{x^3}y\right] &= 0, \\ e^{x^3}y &= c, \\ y &= ce^{-x^3}. \end{aligned}$$

We could also have solved the equation by separating the variables, but the above method is simpler and cleaner.

EXAMPLE 28.2

Solve $\dfrac{dy}{dx} + \dfrac{1}{x}y = \sqrt{x}.$

Solution

Here $p(x) = \frac{1}{x}$, $P(x) = \log x$, and we multiply both sides by $e^{\log x} = x$:

$$\begin{aligned} x\frac{dy}{dx} + y &= x^{\frac{3}{2}}, \\ \frac{d}{dx}[xy] &= x^{\frac{3}{2}}, \\ xy &= \frac{2}{5}x^{\frac{5}{2}} + C, \\ y &= \frac{2}{5}x^{\frac{3}{2}} + Cx^{-1}. \end{aligned}$$

EXAMPLE 28.3

Solve $\dfrac{dy}{dx} + y = x^2$.

Solution

We multiply both sides by e^x:

$$e^x \frac{dy}{dx} + e^x y = x^2 e^x,$$

$$\frac{d}{dx}[e^x y] = x^2 e^x,$$

$$e^x y = \int x^2 e^x \, dx + C.$$

Integrating by parts twice, or more efficiently, consulting an integral table, we get

$$e^x y = x^2 e^x - 2xe^x + 2e^x + C,$$

$$y = x^2 - 2x + 2 + Ce^{-x}.$$

The general formula for the particular solution in Example 28.3 is

$$e^{-x} \int e^x x^2 \, dx = x^2 - 2x + 2. \tag{28.7}$$

For any equation

$$\frac{dy}{dx} + ay = q(x),$$

with $p(x)$ a constant, a, the particular solution will be

$$y_0 = e^{-ax} \int e^{ax} q(x) \, dx. \tag{28.8}$$

If $q(x)$ is a polynomial, then (28.8) will again be a polynomial of the same degree. For instance, in Example 28.3, we had $q(x) = x^2$, and $y_0 = x^2 - 2x + 2$. Rather than fight through the integration in (28.8), we can just substitute a general polynomial in the equation and see what the coefficients must be.

Second Solution to Example 28.3. Since $q(x) = x^2$, we try

$$y = A + Bx + Cx^2,$$

$$\frac{dy}{dx} = B + 2Cx.$$

Substituting in the equation, we get

$$(B + 2Cx) + (A + Bx + Cx^2) = x^2,$$

$$(A + B) + (B + 2C)x + Cx^2 = x^2.$$

Hence, we must have

$$A + B = 0$$

$$B + 2C = 0$$

$$C = 1.$$

Thus, $C = 1$, $B = -2$, $A = 2$, and

$$y_0 = 2 - 2x + x^2$$

is the same as the particular solution given by the integral (28.7).

It is important to emphasize here that the substitution trick works only if the coefficient $p(x)$ is a constant, so the equation looks like this:

$$\frac{dy}{dx} + ay = q(x).$$

However, if $p(x)$ is constant, the substitution technique works not just for polynomials $q(x)$, but also if $q(x) = K \cos bx$ or $q(x) = K \sin bx$ or $q(x) = K_1 \cos bx + K_2 \sin bx$. In this case, try $y = A \cos bx + B \sin bx$. You must include both sine and cosine terms in your trial function, even if $q(x)$ has only one term. If $q(x) = Ke^{bx}$, then try $y = Ae^{bx}$ provided $b \neq -a$. If $b = -a$, e^{bx} is a solution of the reduced equation, and the particular solution has the form $y = Axe^{-ax}$ (Problem 28.14).

EXAMPLE 28.4

Solve $\dfrac{dy}{dx} - 4y = 5 \cos 2x$.

Solution

We know the solutions of the reduced equation are $y = Ce^{4x}$. To find a particular solution, we try

$$y = A \cos 2x + B \sin 2x$$

$$\frac{dy}{dx} = -2A \sin 2x + 2B \cos 2x.$$

Substituting in the equation gives

$$(-2A \sin 2x + 2B \cos 2x) - 4(A \cos 2x + B \sin 2x) = 5 \cos 2x,$$

$$(2B - 4A) \cos 2x + (-2A - 4B) \sin 2x = 5 \cos 2x.$$

Therefore, y will be a solution if

$$-4A + 2B = 5$$

$$-2A - 4B = 0.$$

Multiply the first equation by two and add:

$$(-8A - 2A) + (4B - 4B) = 10,$$

$$-10A = 10,$$

$$A = -1.$$

From the second equation $B = -\frac{1}{2}A$, so $B = \frac{1}{2}$, and the particular solution is

$$y_0 = -\cos 2x + \frac{1}{2} \sin 2x.$$

The general solution is

$$y = -\cos 2x + \frac{1}{2} \sin 2x + Ce^{4x}.$$

EXAMPLE 28.5

$\dfrac{dy}{dx} - y = 4e^{3x}$.

Solution

We try

$$y = Ae^{3x}, \quad \frac{dy}{dx} = 3Ae^{3x}.$$

Substitution yields the condition

$$3Ae^{3x} - Ae^{3x} = 4e^{3x}, \quad 2A = 4, \quad A = 2.$$

The general solution is

$$y = 2e^{3x} + Ce^{x}.$$

EXAMPLE 28.6

$\frac{dy}{dx} + 2y = 5e^{-2x}$.

Here $y = ce^{-2x}$ is the general solution of the reduced equation, and to get a particular solution we try

$$y = Axe^{-2x}, \quad \frac{dy}{dx} = Ae^{-2x} - 2Axe^{-2x}.$$

Substituting these in the equation, we get

$$(Ae^{-2x} - 2Axe^{-2x}) + 2Axe^{-2x} = 5e^{-2x},$$
$$Ae^{-2x} = 5e^{-2x}, \quad A = 5.$$

The general solution is

$$y = 5xe^{-2x} + ce^{-2x}.$$

EXAMPLE 28.7

Recall from the last chapter the differential equation for the velocity v in ft/sec of a falling body with air resistance:

$$\frac{dv}{dt} = 32 - kv,$$
$$\frac{dv}{dt} + kv = 32.$$

The constant right side, 32, is a zero degree polynomial, so there will be a constant particular solution $y = A$. Substituting, we get

$$\frac{dA}{dt} + kA = 32.$$

Since $\frac{dA}{dt} = 0$, $A = \frac{32}{k}$, and the general solution is

$$v = \frac{32}{k} + Ce^{-kt}.$$

With the initial condition $v = 0$ at $t = 0$, this gives $C = -\frac{32}{k}$.

EXAMPLE 28.8

If T is the temperature of a beaker of water at time T and $T = 90°C$ at $t = 0$, $T = 40°C$ at $t = 10$ minutes in a room of temperature 25°C, what is T at $t = 20$ minutes? Assume Newton's law of cooling: $\frac{dT}{dt} = k(T - 25)$.

Solution

We write the equation

$$\frac{dT}{dt} - kT = -25k.$$

The coefficient $p(x)$ is constant, $-k$, and the right side is constant, so there will be a constant particular solution $T = A$. Substituting gives $-kA = -25k$, or $A = 25$. Thus, the solution is

$$T = 25 + Ce^{-kt}.$$

Since $T = 85$ at $t = 0$, $C = 60$ and

$$T = 25 + 60e^{-kt}.$$

Since $T = 40$ at $t = 10$,

$$40 = 25 + 60e^{-10k}$$
$$\frac{15}{60} = (e^{-k})^{10}$$
$$e^{-k} = \left(\frac{1}{4}\right)^{\frac{1}{10}}.$$

For any t

$$T = 25 + 60\left(\frac{1}{4}\right)^{\frac{t}{10}},$$

and when $t = 15$,

$$T = 25 + 60\left(\frac{1}{4}\right)^{\frac{3}{2}} + 25 + 60\left(\frac{1}{8}\right) = 32.5.$$

PROBLEMS

Solve these equations:

28.1 $\dfrac{dy}{dx} + \dfrac{1}{x}y = 0$

28.2 $\dfrac{dy}{dx} + \dfrac{1}{x}y = x$

28.3 $\dfrac{dy}{dx} + 2xy = 0$

28.4 $\dfrac{dy}{dx} + 2xy = x$

28.5 $\dfrac{dy}{dx} + y = e^x$

28.6 $\dfrac{dy}{dx} + y = 2e^{-x}$ (Try $y = Axe^{-x}$.)

28.7 $\dfrac{dy}{dx} - 3y = 6x + 1$

28.8 $\dfrac{dy}{dx} + 2y = 4x^2 + 4x$

28.9 $\dfrac{dy}{dx} - 6y = -15\sin 3x$

28.10 $\dfrac{dy}{dx} + y = \cos 2x$

28.11 $\dfrac{dy}{dx} + y = e^x + \cos 2x$ *Hint*: Look at Problems 28.5 and 28.10, and the linearity condition (28.2).

28.12 $\dfrac{dy}{dx} - 6y = 6x - 15\sin 3x$ *Hint*: Use Problem 28.9.

28.13 $\dfrac{dy}{dx} + 3y = 5e^{2x}$

28.14 Show that $\frac{dy}{dx} + ay = Ke^{bx}$ has a solution $y = Ae^{bx}$ unless $b = -a$. If $b = -a$, there is a solution of the form $y = Axe^{-ax}$.

28.15 Suppose a light body falls under gravity, with air resistance as in Example 28.7, and its terminal velocity is 16 ft/sec. How fast is it falling after one second?

28.16 If $T = 25 + Ce^{-kt}$ as in Example 28.8, and $T = 75$ when $t = 0$ and $T = 65$ when $t = 5$ min, what is the temperature at 15 min?

29

Homogeneous Second-Order Linear Equations

The two common physical situations indicated schematically in Figures 29.1 and 29.2 are typical applications of the kind of equation we study in this chapter. In Figure 29.1 we show a mass m attached to a spring and a damping device that could simply represent friction. The spring exerts a force $-ks$ on the mass, where s is the amount the spring stretched ($s > 0$) or compressed ($s < 0$), and the damper exerts on the mass a retarding force $-c\frac{ds}{dt}$ proportional to the speed. Newton's law says that force is mass times acceleration, so we have the equation of motion

$$m\frac{d^2s}{dt^2} = -ks - c\frac{ds}{dt},$$

or

$$\frac{d^2s}{dt^2} + \frac{c}{m}\frac{ds}{dt} + \frac{k}{m}s = 0. \tag{29.1}$$

Figure 29.2 represents a simple LCR circuit with an inductance L, a capacitance C, and a resistance R. The current I in the circuit is determined by

$$\frac{d^2I}{dt^2} + \frac{R}{L}\frac{dI}{dt} + \frac{1}{LC}I = 0. \tag{29.2}$$

Both equations (29.1) and (29.2) have the same form,

$$\frac{d^2y}{dt^2} + b\frac{dy}{dt} + cy = 0, \tag{29.3}$$

where b and c are constants. Equation (29.3) is the general form of a **homogeneous linear second-order equation with constant coefficients**.

If we denote the left side of (29.3) by $L(y)$; that is, let

$$L(y) = \frac{d^2y}{dt^2} + b\frac{dy}{dt} + cy, \tag{29.4}$$

then the linearity properties of L are

$$L(ky) = kL(y) \text{ and } L(y_1 + y_2) = L(y_1) + L(y_2) \tag{29.5}$$

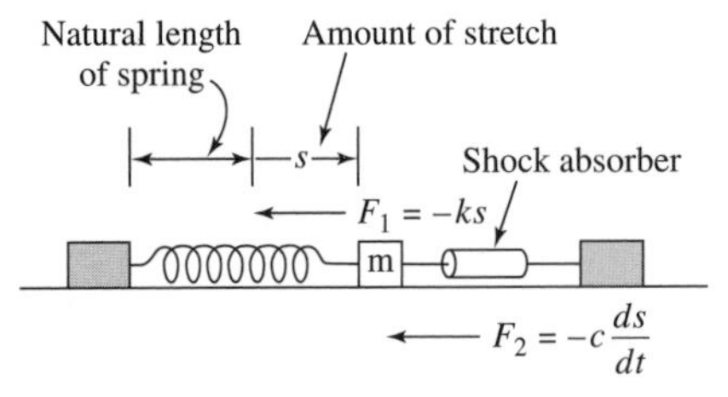

Figure 29.1

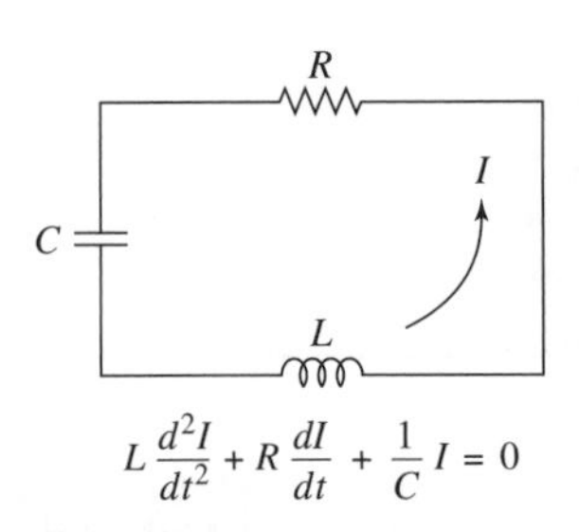

Figure 29.2

for every constant k and all functions y, y_1, y_2. From (29.5) we see that if y_1 and y_2 are solutions of (29.3), so that $L(y_1) = 0$ and $L(y_2) = 0$, then any linear combination $C_1y_1 + C_2y_2$ is also a solution. In fact, the general solution of (29.3) is

$$y = C_1y_1 + C_2y_2, \tag{29.6}$$

provided y_1 and y_2 are **independent** solutions of (29.3); that is, neither y_1 nor y_2 is a multiple of the other. To determine a specific function from the two-parameter family (29.6), we specify an **initial condition** of the form $y(t_0) = a_0$, $y'(a_0) = a_1$; that is, we must specify the values of both y and $\frac{dy}{dt}$ at some initial time t_0.

In the simple case $\frac{d^2y}{dt^2} = 0$, we have already seen that the solutions are $y = C_1 + C_2t$. If either b or c is not zero in the general equation (29.3), then we look for solutions of the form $y = e^{mt}$. Substituting $y = e^{mt}$, $\frac{dy}{dt} = me^{mt}$, $\frac{d^2y}{dt^2} = m^2e^{mt}$ in (29.3) we get

$$e^{mt}(m^2 + bm + c) = 0.$$

Thus, if r is a root of the **auxiliary equation**

$$m^2 + bm + c = 0, \tag{29.7}$$

then $y = e^{rt}$ is a solution of the differential equation. If the auxiliary equation has two real roots r_1 and r_2, then e^{r_1t} and e^{r_2t} are independent solutions, and the general solution is

$$y = C_1e^{r_1t} + C_2e^{r_2t}. \tag{29.8}$$

EXAMPLE 29.1

Solve $\frac{d^2y}{dt^2} - 3\frac{dy}{dt} + 2y = 0$, and find the solution such that $y(0) = 3$, $y'(0) = 5$.

Solution

The auxiliary equation is

$$m^2 - 3m + 2 = (m - 1)(m - 2) = 0,$$

with roots 1 and 2. The general solution is

$$y = C_1e^t + C_2e^{2t}.$$

Now use $y(0) = 3$, $y'(0) = 5$ to find C_1 and C_2. Since $y'(t) = C_1e^t + 2C_2e^{2t}$, at $t_0 = 0$ we have:

$$C_1 + C_2 = 3$$
$$C_1 + 2C_2 = 5.$$

Subtracting the first equation from the second gives $C_2 = 2$, and then from the first equation we get $C_1 = 1$. The solution that satisfies the initial condition is

$$y = e^t + 2e^{2t}.$$

If the auxiliary equation has only one real root, r, then it is easy to check (Problem 29.6) that e^{rt} and te^{rt} are two independent solutions of the differential equation.

EXAMPLE 29.2

Solve $\dfrac{d^2y}{dt^2} + 4\dfrac{dy}{dt} + 4y = 0.$

Solution

Here the auxiliary equation is

$$m^2 + 4m + 4 = (m+2)^2 = 0,$$

with the single root $r = -2$. The functions e^{-2t} and te^{-2t} are independent solutions, and the general solution is

$$y = C_1 e^{-2t} + C_2 t e^{-2t}.$$

The roots of the auxiliary equation for the general equation (29.3) are

$$\frac{-b \pm \sqrt{b^2 - 4c}}{2} = -\frac{b}{2} \pm \sqrt{\left(\frac{b}{2}\right)^2 - c}. \tag{29.9}$$

If b and c are both positive, as they are in the physical situations pictured in Figures 29.1 and 29.2, then both the roots r_1 and r_2 in (29.9) will be negative, so $e^{r_1 t}$ and $e^{r_2 t}$ both approach zero as $t \longrightarrow \infty$. This is an obvious physical necessity, since neither system has any external source of energy and all solutions are necessarily transients.

If the auxiliary equation has complex roots $r + is$ and $r - is$, then $y_1 = e^{(r+is)t}$ and $y_2 = e^{(r-is)t}$ are complex valued solutions of the differential equation. The famous **Euler formula** states that

$$e^{(r \pm is)t} = e^{rt} \cos st \pm i e^{rt} \sin st.$$

Both the real and imaginary parts of this complex function will be real solutions to the differential equation. That is, if $r \pm is$ are complex roots of the auxiliary equation, then $y_1 = e^{rt} \cos st$ and $y_2 = e^{rt} \sin st$ are independent solutions to the differential equation, and the general solution is

$$y = C_1 e^{rt} \cos st + C_2 e^{rt} \sin st.$$

EXAMPLE 29.3

Solve $\dfrac{d^2y}{dt^2} + 4\dfrac{dy}{dt} + 13y = 0.$

Solution

The auxiliary equation is

$$m^2 + 4m + 13 = 0,$$

and the quadratic formula gives the roots

$$\begin{aligned} r \pm is &= \frac{-4 \pm \sqrt{16 - 4 \cdot 13}}{2} \\ &= -2 \pm \sqrt{4 - 13} = -2 \pm 3i. \end{aligned}$$

The solutions are

$$y = C_1 e^{-2t} \cos 3t + C_2 e^{-2t} \sin 3t.$$

EXAMPLE 29.4

Solve $\dfrac{d^2y}{dt^2} + 9y = 0$.

Solution

The two roots of $m^2 + 9 = 0$ are $3i$ and $-3i$. Since $r = 0$ and $e^{0t} = 1$, the two solutions are $\cos 3t$ and $\sin 3t$, and the general solution is

$$y = C_1 \cos 3t + C_2 \sin 3t.$$

Any equation of the form

$$\frac{d^2y}{dt^2} + \omega^2 y = 0,$$

as in Example 29.4, describes what is called **simple harmonic motion**. The general solution is

$$y = C_1 \cos \omega t + C_2 \sin \omega t. \tag{29.10}$$

By rewriting (29.10) as follows,

$$y = \sqrt{C_1^2 + C_2^2}\left[\frac{C_1}{\sqrt{C_1^2 + C_2^2}} \cos \omega t + \frac{C_2}{\sqrt{C_1^2 + C_2^2}} \sin \omega t\right],$$

letting $K = \sqrt{C_1^2 + C_2^2}$, and defining α by

$$\sin \alpha = \frac{C_1}{\sqrt{C_1^2 + C_2^2}}, \qquad \cos \alpha = \frac{C_2}{\sqrt{C_1^2 + C_2^2}},$$

we can write the general solution (29.10) in the form

$$y = K \sin(\omega t + \alpha). \tag{29.11}$$

This formula exhibits the simple harmonic motion as a sine wave with **amplitude** K and **frequency** $\frac{\omega}{2\pi}$. The number α is called the **phase shift**.

EXAMPLE 29.5

The general solution of the simple harmonic motion $\frac{d^2y}{dt^2} + 9y = 0$ can be written either as $y = C_1 \cos 3t + C_2 \sin 3t$ as in Example 29.4 or as $y = K \sin(3t + \alpha)$. Find K and α so that $y(0) = 1$, $y'(0) = 3$. What is the frequency?

Solution

We have

$$y(t) = K \sin(3t + \alpha), \quad y(0) = K \sin \alpha,$$
$$y'(t) = 3K \cos(3t + \alpha), \quad y'(0) = 3K \cos \alpha.$$

Therefore, $\sin \alpha = \frac{y(0)}{K}$ and $\cos \alpha = \frac{y'(0)}{3K}$, so

$$\frac{\sin \alpha}{\cos \alpha} = \frac{3y(0)}{y'(0)} = \frac{3}{3} = 1,$$
$$\alpha = \frac{\pi}{4}.$$

Hence,

$$y(0) = 1 = K \sin \frac{\pi}{4} = K \frac{1}{\sqrt{2}},$$
$$K = \sqrt{2}.$$

Hence, the amplitude, K, is $\sqrt{2}$, and the phase shift, α, is $\frac{\pi}{4}$. The frequency is $\frac{3}{2\pi}$. The solution is

$$y = \sqrt{2}\sin\left(3t + \frac{\pi}{4}\right).$$

PROBLEMS

29.1 Verify that $L(y)$, defined in (29.4), has the linearity properties (29.5).

Find the general solutions, as well as the specific solution that satisfies the initial condition.

29.2 $\dfrac{d^2y}{dt^2} + 5\dfrac{dy}{dt} + 6y = 0;\quad y(0) = 1,\ y'(0) = 5$

29.3 $\dfrac{d^2y}{dt^2} - \dfrac{dy}{dt} - 2y = 0;\quad y(0) = 2,\ y'(0) = 7$

29.4 $\dfrac{d^2y}{dt^2} + 3\dfrac{dy}{dt} + 2y = 0;\quad y(0) = -1,\ y'(0) = 1$

29.5 $\dfrac{d^2y}{dt^2} - 2\dfrac{dy}{dt} - 3y = 0;\quad y(1) = 2e^3 + e^{-1},\ y'(1) = 6e^3 - e^{-1}$

29.6 Suppose the auxiliary equation has just one root, r, so the equation is $(m - r)^2 = 0$. Write the differential equation and verify that te^{rt} is a solution.

29.7 Find the general solution of $\frac{d^2y}{dt^2} - 2\frac{dy}{dt} + y = 0$.

29.8 Find the general solution of $\frac{d^2y}{dt^2} + 10\frac{dy}{dt} + 25y = 0$.

29.9 Show that the auxiliary equation for $\frac{d^2y}{dt^2} - 2A\frac{dy}{dt} + (A^2 + B^2)y = 0$ has complex roots $A \pm iB$. Check that $y_1 = e^{At}\cos Bt$ and $y_2 = e^{At}\sin Bt$ are independent solutions.

Find the general solution and the specific solution that satisfies the given initial conditions.

29.10 $\dfrac{d^2y}{dt^2} - 2\dfrac{dy}{dt} + 2y = 0;\quad y(0) = 3,\ y'(0) = 10$

29.11 $\dfrac{d^2y}{dt^2} - 2\dfrac{dy}{dt} + 5y = 0;\quad y(0) = 3,\ y'(0) = 11$

29.12 $\dfrac{d^2y}{dt^2} - 6\dfrac{dy}{dt} + 10y = 0;\quad y(0) = 0,\ y'(0) = 6$

29.13 $\dfrac{d^2y}{dt^2} - 4\dfrac{dy}{dt} + 20y = 0;\quad y\left(\dfrac{\pi}{4}\right) = -e^{\frac{\pi}{2}},\ y'\left(\dfrac{\pi}{4}\right) = -2e^{\frac{\pi}{2}}$

29.14 Solve $\frac{d^2y}{dt^2} + 4y = 0$ and find the frequency of this simple harmonic motion.

29.15 Write the solution of $\frac{d^2y}{dt^2} + 4y = 0$ in the form $y = K\sin(\omega t + \alpha)$. What is ω here? Find K and α so that $y(0) = \sqrt{3}$, $y'(0) = 2$.

29.16 If a pendulum has length ℓ feet and θ is the angle the pendulum has swung from the vertical at time t, then $\frac{d^2\theta}{dt^2} + \frac{g}{\ell}\theta = 0$, where $g = 32$ ft/sec^2 is the acceleration of gravity.

(a) How long should ℓ be so that the pendulum swings from one side to the other in one second? *Hint*: One cycle would consist of a swing from the extreme right to the extreme left and back to the extreme right. The frequency in this situation is therefore $\frac{1}{2}$ cycle per second.

(b) With ℓ the length of part (a), suppose $\theta = 0$ when $t = 0$ (so the phase shift α is zero), and $\frac{d\theta}{dt} = .495$ (radians per second, or about 28 degrees/sec) when $t = 0$. Find θ in terms of t. What is the maximum angle θ the pendulum makes with the vertical?

30

Nonhomogeneous Second-Order Equations

In this chapter, we treat the nonhomogeneous linear second-order equation

$$\frac{d^2y}{dt^2} + a\frac{dy}{dt} + by = q(t), \tag{30.1}$$

where a and b are still constants but $q(t)$ is an arbitrary function. In physical terms, $q(t)$ could be an imposed electromotive force in the LCR circuit of the last chapter or an outside force acting on the spring-mass system.

The **reduced form** of (30.1) is the homogeneous equation

$$\frac{d^2y}{dt^2} + a\frac{dy}{dt} + by = 0. \tag{30.2}$$

If we let $L(y)$ denote the left side of (30.1) or (30.2), then $L(y)$ is a **linear operator**, which means that

$$L(ky) = kL(y) \text{ and } L(y_1 + y_2) = L(y_1) + L(y_2) \tag{30.3}$$

for all constants k and functions y, y_1, y_2. From the linearity properties (30.3) it is clear that if y_0 is any solution of (30.1), so $L(y_0) = q(t)$, and y_1, y_2 are solutions of (30.2), so $L(y_1) = L(y_2) = 0$, then

$$y = y_0 + C_1y_1 + C_2y_2 \tag{30.4}$$

is a solution of (30.1), since

$$\begin{aligned} L(y_0 + C_1y_1 + C_2y_2) &= L(y_0) + C_1L(y_1) + C_2L(y_2) \\ &= q(t) + C_1 \cdot 0 + C_2 \cdot 0. \end{aligned}$$

In fact, (30.4) is the general solution of (30.1) provided y_1 and y_2 are **independent** solutions of (30.2); that is, neither is a multiple of the other.

Recall that for the first-order equation

$$\frac{dy}{dx} + p(x)y = q(x) \tag{30.5}$$

we found the following general solution:

$$y = e^{-P(x)} \int e^{P(x)} q(x)\, dx + Ce^{-P(x)}, \tag{30.6}$$

where $P(x) = \int p(x)\, dx$, and $e^{-P(x)}$ is a solution of the reduced form of (30.5). From (30.6) we see that (30.5) has a particular solution of the form $y = y_0 v$, where $y_0 = e^{-P(x)}$ is a solution of the reduced equation, and v is the function given by the integral in (30.6). With this guideline, we try to find a particular solution of (30.1) of the form $y = y_0 v$, where y_0 is a solution of the reduced equation. So let $y_0 = e^{rt}$ be a solution of the reduced equation (30.2), and substitute $y = e^{rt} v$ into (30.1) to see what v must be if y is to be a solution. The terms involving v drop out because $L(e^{rt}) = 0$, and we get

$$\begin{aligned} e^{rt}\frac{d^2v}{dt^2} + (2r+a)e^{rt}\frac{dv}{dt} &= q(t), \\ \frac{d^2v}{dt^2} + (2r+a)\frac{dv}{dt} &= e^{-rt}q(t). \end{aligned} \tag{30.7}$$

Equation (30.7) is a first-order equation in $\frac{dv}{dt}$; let $u = \frac{dv}{dt}$ and (30.7) becomes

$$\frac{du}{dt} + (2r+a)u = e^{-rt}q(t). \tag{30.8}$$

We solve this as usual by multiplying both sides by $e^{P(t)}$ where $P(t) = \int p(t)\, dt = (2r+a)t$.

$$\begin{aligned} e^{(2r+a)t}\frac{du}{dt} + (2r+a)e^{(2r+a)t}u &= e^{(2r+a)t}e^{-rt}q(t), \\ \frac{d}{dt}(e^{(2r+a)t}u) &= e^{(r+a)t}q(t), \\ e^{(2r+a)t}u &= \int e^{(r+a)t}q(t)\, dt. \end{aligned} \tag{30.9}$$

Now we have an integral formula for u:

$$u = e^{-(2r+a)t}\int e^{(r+a)t}q(t)\, dt. \tag{30.10}$$

Since $v = \int u(t)\, dt$, we also have an integral formula for v, and therefore a guaranteed solution $y = e^{rt}v$, where v is given explicitly in terms of two integrations involving $q(t)$ and exponentials. We chase through the computations above in the following example.

EXAMPLE 30.1

Solve $\dfrac{d^2y}{dt^2} - 3\dfrac{dy}{dt} + 2y = 4t$.

Solution

The reduced equation has solutions e^t and e^{2t}. We let $y_0 = e^t$ and substitute $y = e^t v$ in the equation to get

$$\begin{aligned} e^t\frac{d^2v}{dt^2} - e^t\frac{dv}{dt} &= 4t, \\ \frac{d^2v}{dt^2} - \frac{dv}{dt} &= 4te^{-t}. \end{aligned}$$

Notice that v is missing from this equation as in (30.7). We let $u = \frac{dv}{dt}$ and multiply both sides by e^{-t}:

$$\begin{aligned} e^{-t}\frac{du}{dt} - e^{-t}u &= 4te^{-2t}, \\ e^{-t}u &= \int 4te^{-2t}\, dt. \end{aligned}$$

Integrate by parts (or consult the table of integrals) to get

$$e^{-t}u = e^{-2t}(-2t-1),$$
$$u = e^{-t}(-2t-1).$$

Now integrate u to get v:

$$v = \int e^{-t}(-2t-1)\,dt = e^{-t}(2t+3).$$

This gives the particular solution

$$y = e^{t}v = e^{t}\cdot e^{-t}(2t+3) = 2t+3.$$

Notice that although $q(t) = t$ is a very simple function, the computations are nevertheless formidable. This prompts us to investigate what general form a solution would have for certain functions $q(t)$. For example, if we knew that a polynomial $q(t)$ would always lead to a polynomial solution (and it does), then it would be much easier just to substitute a general polynomial $At + B$ in the equation to determine the coefficients A and B.

Suppose $q(t) = e^{kt}Q_n(t)$, where $Q_n(t)$ is an nth degree polynomial. Here are some examples of this form of $q(t)$:

$$q(t) = 5e^{3t}; \quad Q_0(t) = 5,\ k = 3;$$
$$q(t) = 3t^3 + t; \quad Q_3(t) = 3t^3 + t,\ k = 0 \text{ so } e^{kt} = 1;$$
$$q(t) = 2te^{-t}; \quad Q_1(t) = 2t,\ k = -1.$$

If $q(t) = e^{kt}Q_n(t)$, then the integral in (30.10) has the form

$$\int e^{(r+a)t}e^{kt}Q_n(t)\,dt = \int e^{ct}Q_n(t)\,dt, \tag{30.11}$$

where $c = r + a + k$ is a constant. From the integral tables (or see Problem 30.1), we find that if $c \neq 0$, then

$$\int e^{ct}Q_n(t)\,dt = e^{ct}R_n(t), \tag{30.12}$$

where $R_n(t)$ is another nth degree polynomial. Hence (30.9) looks like this:

$$\begin{aligned} e^{(2r+a)t}u &= \int e^{(r+a+k)t}Q_n(t)\,dt \\ &= e^{(r+a+k)t}R_n(t), \end{aligned} \tag{30.13}$$

from which

$$u = e^{(k-r)t}R_n(t). \tag{30.14}$$

The function v is the integral of u, so if $k - r \neq 0$,

$$v = \int e^{(k-r)t}R_n(t)\,dt = e^{(k-r)t}S_n(t), \tag{30.15}$$

for some new nth degree polynomial $S_n(t)$. If $k = r$, so e^{kt} is a solution of the reduced equation, then

$$v = \int R_n(t)\,dt = T_{n+1}(t), \tag{30.16}$$

where $T_{n+1}(t)$ is a $(n+1)$st degree polynomial with no constant term. We can write $T_{n+1}(t)$ in the form $tS_n(t)$.

Finally, the solution $y = e^{rt}v$ has the form

$$y = e^{rt}v = e^{rt}e^{(k-r)t}S_n(t) = e^{kt}S_n(t) \tag{30.17}$$

if $k \neq r$, and has the form

$$y = e^{kt} t S_n(t)$$

if $k = r$.

To find a particular solution of (30.1) if $q(t) = e^{kt} Q_n(t)$, *try* $y = e^{kt} S_n(t)$ *if* e^{kt} *is not a solution of the reduced equation, and try* $y = e^{kt} t S_n(t)$ *if* e^{kt} *is a solution of the reduced equation.*

Here are some examples of this rule. In all these examples, the reduced equation is $\frac{d^2y}{dt^2} - 3\frac{dy}{dt} + 2y = 0$, with solutions e^t and e^{2t}.

$$\text{(a)} \quad \frac{d^2y}{dt^2} - 3\frac{dy}{dt} + 2y = 6e^{-t}.$$

There is a particular solution of the form $y = Ae^{-t}$.

$$\text{(b)} \quad \frac{d^2y}{dt^2} - 3\frac{dy}{dt} + 2y = e^t.$$

Since e^t is a solution of the reduced equation, there will be a solution of the form Ate^t.

$$\text{(c)} \quad \frac{d^2y}{dt^2} - 3\frac{dy}{dt} + 2y = 4t^2e^{3t}.$$

There is a solution of the form $y = (At^2 + Bt + C)e^{3t}$.

$$\text{(d)} \quad \frac{d^2y}{dt^2} - 3\frac{dy}{dt} + 2y = (-6t + 8)e^{2t}.$$

There will be a solution of the form $y = (At^2 + Bt)e^{2t}$ since e^{2t} is a solution of the reduced equation.

$$\text{(e)} \quad \frac{d^2y}{dt^2} - 3\frac{dy}{dt} + 2y = 4t.$$

This is the problem of Example 30.1. Now we know there is a solution of the form $y = At + B$, so we can substitute this general linear polynomial and determine A and B. That's a whole lot simpler than the calculations of Example 30.1.

The same kind of argument we used above shows that if $q(t) = Q_n(t) \cos \alpha t$ or $q(t) = Q_n(t) \sin \alpha t$ for a polynomial $Q_n(t)$, then there will be a solution of the form

$$y = R_n(t) \cos \alpha t + S_n(t) \sin \alpha t.$$

Notice that both $\cos \alpha t$ and $\sin \alpha t$ must occur in the trial solution, even if $q(t)$ involves one of these functions. For example,

$$\text{if } q(t) = 3 \sin 4t, \text{ try } y = A \cos 4t + B \sin 4t;$$
$$\text{if } q(t) = 2t \cos t,$$
$$\text{try } y = (At + B) \cos t + (Ct + D) \sin t.$$

If $\cos \alpha t$ and $\sin \alpha t$ are solutions of the reduced equation, both $R_n(t)$ and $S_n(t)$ must be multiplied by t.

EXAMPLE 30.2

Solve $\dfrac{d^2y}{dt^2} - \dfrac{dy}{dt} - 2y = 10 \cos t.$

Solution

The auxiliary equation is

$$m^2 - m - 2 = (m-2)(m+1) = 0,$$

so e^{2t} and e^{-t} are solutions of the reduced equation. There will be a particular solution of the form $y = A\cos t + B\sin t$, so we calculate derivatives and substitute:

$$\begin{aligned} y &= A\cos t + B\sin t \\ \frac{dy}{dt} &= B\cos t - A\sin t \\ \frac{d^2y}{dt^2} &= -A\cos t - B\sin t. \end{aligned}$$

Substituting in the equation gives

$$\begin{aligned} (\cos t)(-A - B - 2A) + (\sin t)(-B + A - 2B) &= 10\cos t, \\ (-3A - B)\cos t + (A - 3B)\sin t &= 10\cos t. \end{aligned}$$

Hence, we must have

$$\begin{aligned} -3A - B &= 10 \\ A - 3B &= 0. \end{aligned}$$

Multiply the first equation by -3 and add:

$$10A = -30, \quad A = -3.$$

This gives $B = -3A - 10 = -1$, and hence

$$y = -3\cos t - \sin t.$$

EXAMPLE 30.3

Solve $\dfrac{d^2y}{dt^2} + 4y = 4\sin 2t$.

Solution

Here $\cos 2t$ and $\sin 2t$ are solutions of the reduced equation, so the trial solution has the form

$$\begin{aligned} y &= At\cos 2t + Bt\sin 2t, \\ \frac{dy}{dt} &= A\cos 2t - 2At\sin 2t + B\sin 2t + 2Bt\cos 2t \\ &= (\cos 2t)(A + 2Bt) + (\sin 2t)(-2At + B) \\ \frac{d^2y}{dt^2} &= (\cos 2t)(2B) - 2(A + 2Bt)\sin 2t \\ &\quad +(\sin 2t)(-2A) + 2(-2At + B)\cos 2t \\ &= (\cos 2t)(-4At + 4B) + \sin 2t(-4Bt - 4A). \end{aligned}$$

Substitution in the equation gives

$$(\cos 2t)[4B] + (\sin 2t)[-4A] = 4\sin 2t.$$

Hence,

$$\begin{aligned} -4A &= 4, \quad A = -1 \\ 4B &= 0, \quad B = 0 \end{aligned}$$

and the solution is

$$y = -t\cos 2t.$$

PROBLEMS

30.1 (a) Integrate by parts to show that if $n \geq 1$,

$$\int t^n e^{ct}\, dt = \frac{1}{c} t^n e^{ct} - \frac{n}{c} \int t^{n-1} e^{ct}\, dt$$

(b) Use (a) and $\int e^{ct}\, dt = \frac{1}{c} e^{ct}$ to integrate the following in turn:

$$\int t e^{ct}\, dt, \quad \int t^2 e^{ct}\, dt, \quad \int t^3 e^{ct}\, dt.$$

(c) Show why $\int e^{ct} Q_n(t)\, dt = e^{ct} R_n(t)$ for some nth degree polynomial $R_n(t)$ if $Q_n(t)$ is an nth degree polynomial and $c \neq 0$.

Find a particular solution using the examples (a)–(e) in the text.

30.2 $\dfrac{d^2y}{dt^2} - 3\dfrac{dy}{dt} + 2y = 6e^{-t}$ (Example (a))

30.3 $\dfrac{d^2y}{dt^2} - 3\dfrac{dy}{dt} + 2y = e^{t}$ (Example (b))

30.4 $\dfrac{d^2y}{dt^2} - 3\dfrac{dy}{dt} + 2y = 4t^2 e^{3t}$ (Example (c))

30.5 $\dfrac{d^2y}{dt^2} - 3\dfrac{dy}{dt} + 2y = (-6t + 8)e^{2t}$ (Example (d))

30.6 $\dfrac{d^2y}{dt^2} - 3\dfrac{dy}{dt} + 2y = 4t$ (Example (e))

30.7 Solve $\frac{d^2y}{dt^2} - \frac{dy}{dt} - 2y = 10 \sin t$.

30.8 Solve $\frac{d^2y}{dt^2} + y = 6 \cos t$.

30.9 Solve $\frac{d^2y}{dt^2} - 2\frac{dy}{dt} + y = -25 \sin 2t$.

30.10 Solve $\frac{d^2y}{dt^2} - 2\frac{dy}{dt} + 2y = e^t \cos t$. *Hint*: Try $y = Ate^t \cos t + Bte^t \sin t$.

Answers

CHAPTER 1

1.1 (a, b) and (b, a) are symmetric about the line $y = x$.
1.2 $m = -\frac{1}{2}$
1.3 $m = 1$
1.4 $m = 2$
1.5 $m = -1$
1.6 $m = -\frac{1}{3}$
1.7 $m = -\frac{2}{3}$
1.8 $y = -\frac{1}{2}x + \frac{3}{2}$
1.9 $y = -\frac{1}{3}x + \frac{7}{3}$
1.10 $y = 2x + 7$
1.11 $y = -4x - 1$
1.12 $y = -3x + 7$
1.13 $y = -2x + 7$
1.14 $y = 7x - 5$
1.15 (iii) $k = \frac{1}{3}$
1.16 $x - 2y + 2 + \frac{1}{2}(x + y - 4) = 0$, or $x - y = 0$
1.17 $y = 2x - 2$
1.18 $-3x + y + 8 = 0$
1.19 8.1°
1.20 $y - 4 = .466(x - 3)$

CHAPTER 2

2.1 Downward parabola with axis $x = 0$, vertex $(0, 0)$
2.2 Upward parabola with axis $x = 0$, vertex $(0, 0)$

2.3 Upward parabola with axis $x = 0$, vertex $(0, 0)$
2.4 Upward parabola with axis $x = -1$, vertex $(-1, 0)$
2.5 Upward parabola with axis $x = 0$, vertex $(0, 1)$
2.6 Downward parabola with axis $x = 0$, vertex $(0, 1)$
2.7 Upward parabola with axis $x = 1$, vertex $(1, 1)$
2.8 Downward parabola with axis $x = 1$, vertex $(1, 1)$
2.9 Circle with center $(0, 0)$, radius 2
2.10 Circle with center $(1, 0)$, radius 1
2.11 Circle with center $(0, 2)$, radius 3
2.12 Ellipse through $(\pm 3, 0)$ and $(0, \pm 2)$
2.13 Ellipse through $(\pm 1, 0)$ and $(0, \pm 2)$
2.14 Ellipse through $(\pm 3, 0)$ and $(0, \pm 2)$
2.15 Hyperbola through $(\pm 1, 0)$, asymptotes $y = \pm x$
2.16 Hyperbola through $(0, \pm 1)$, asymptotes $y = \pm x$
2.17 Hyperbola through $(\sqrt{2}, \sqrt{2})$ and $(-\sqrt{2}, -\sqrt{2})$, asymptotes $x = 0$ and $y = 0$
2.18 Hyperbola through $(2, 1)$ and $(0, -1)$, asymptotes $y = 0$ and $x = 1$
2.19 $x^2 + y^2 - 2x + 4y = 0$
2.20 $y = \frac{1}{3}(x + 1)^2$

CHAPTER 3

3.1 4
3.2 20
3.3 27
3.4 6
3.5 $\frac{1}{2}$
3.6 $\frac{1}{2}$
3.7 $-\frac{1}{9}$
3.8 -2
3.9 $\frac{1}{12}$
3.10 $y - 2 = 4(x - 1)$
3.11 $y - 1 = -4(x - 2)$
3.12 $y - 4 = \frac{1}{2}(x - 4)$
3.13 $y - 1 = -\frac{1}{2}(x - 1)$
3.14 $s'(2) = 36$ ft/sec; $s'(0) = 100$ ft/sec; max height $= 156.25$ ft
3.15 $s(2) = 104$; $s'(2) = 84$ ft/sec
3.16 $\frac{dv}{dt} = 4\pi t^2$
3.17 $W'(100) = -\frac{1}{10}$ lb/min

CHAPTER 4

4.1 $\frac{dy}{dx} = 20x^3 + 2x + 1$
4.2 $\frac{dy}{dx} = -6x + \frac{1}{2}$

4.3 $\frac{dy}{dx} = -12x^{-5} + x^{-2} + 6x^2$
4.4 $\frac{dy}{dx} = -8x^{-5} + 9x^{-4} + 5 + 3x^5$
4.5 $\frac{dy}{dx} = 3x^2 - 3x^{-4}$
4.6 $\frac{dy}{dx} = 10x^9 + 10x^{-11}$
4.7 $\frac{dy}{dx} = 6x^2 + 6x + 2$
4.8 $\frac{dy}{dx} = 4x^3$
4.9 $\frac{dy}{dx} = 10x + 2 + 3x^{-2}$
4.10 $\frac{dy}{dx} = \frac{-4x}{(x^2-1)^2}$
4.11 $\frac{dy}{dx} = \frac{-3x^2}{(2+x^3)^2}$
4.12 $\frac{dy}{dx} = \frac{2x^3+3x^2}{(x+1)^2}$
4.13 $\frac{dy}{dx} = \frac{(2x^4+x^2+1)}{(2x^2+1)^2}$
4.14 $\frac{dy}{dx} = \frac{-4}{(2x-1)^2}$
4.15 $\frac{dy}{dx} = \frac{5}{2\sqrt{x}} - \frac{1}{x^2}$
4.16 $\frac{dy}{dx} = \frac{9}{2}\sqrt{x} - \frac{1}{\sqrt{x}}$
4.17 $\frac{dy}{dx} = -\frac{1}{2x^{\frac{3}{2}}}$
4.20 $\frac{dy}{dx} = (x^2+2)(x-2) + (x+1)(2x)(x-2) + (x+1)(x^2+2)$
4.21 $\frac{dy}{dx} = -3x^{-4}(x+3)(x^2-4) + x^{-3}(x^2-4) + x^{-3}(x+3)(2x)$

CHAPTER 5

5.1 $\frac{dy}{dx} = 12(1+3x)^3$
5.2 $\frac{dy}{dx} = 5(x+x^2)^4(1+2x)$
5.3 $\frac{dy}{dx} = -2(1+2x)^{-2}$
5.4 $\frac{dy}{dx} = -10(x^3+x^7)^{-11}(3x^2+7x^6)$
5.5 $\frac{dy}{dx} = -3(2+3x)^{-2}$
5.6 $\frac{dy}{dx} = -2x(1+x^2)^{-2}$
5.7 $\frac{dy}{dx} = x(1+x^2)^{-\frac{1}{2}}$
5.8 $\frac{dy}{dx} = \frac{3(\sqrt{x}+1)^2}{2\sqrt{x}}$
5.9 $\frac{dy}{dx} = \frac{-6(2x+1)^2(x^2+x-1)}{(1+x^2)^3}$
5.10 $\frac{dy}{dx} = \frac{5x^2(1-x)}{2\sqrt{x}(x+1)^6}$
5.11 $\frac{dy}{dx} = -\frac{3}{2}x^2(x^3-2)^{-\frac{3}{2}}$
5.12 $\frac{dy}{dx} = -2(\sqrt{1+x}+x)^{-3}(\frac{1}{2\sqrt{1+x}}+1)$
5.13 $\frac{dz}{dx} = -6(2x+1)^{-4}$
5.14 $\frac{dz}{dx} = -5(x+1)^{-6}$
5.15 $\frac{dz}{dx} = -\frac{1}{2}x^{-\frac{3}{2}}$
5.16 $\frac{dz}{dx} = -\frac{1}{2}x^{-\frac{3}{2}}$
5.17 $\frac{dz}{dx} = -x(x^2+9)^{-\frac{3}{2}}$

5.18 $\frac{dz}{dx} = \frac{(1+\sqrt{x+1})}{\sqrt{x+1}}$

5.19 $\frac{dA}{dt} = 17\ \text{cm}^2/\text{sec}$

5.20 $\frac{dh}{dt} = \frac{1}{3\pi}$ ft/min; $\frac{dA}{dt} = 1\ \text{ft}^2/\text{min}$

5.21 $\frac{dT}{dt} = \frac{5}{4\sqrt{3}}$ degrees/sec

5.22 $\frac{dh}{dt} = \frac{5}{18\pi}$ in./sec

5.23 $\frac{dr}{dt} = \frac{3}{2^5 \cdot \pi^2 \cdot 5^5} = \frac{3}{10^5 \pi^2}$

5.24 $\frac{50}{\sqrt{5}}$

5.25 $\frac{dy}{dx} = \frac{-x}{\sqrt{a^2-x^2}}$

5.26 $t = 2$

5.27 $g'(x) = (2x+1)/(x^2+x)$

CHAPTER 6

6.3 $\frac{dy}{dx} = -2\sin 2x$

6.4 $\frac{dy}{dx} = 3\cos(3x+1)$

6.5 $\frac{dy}{dx} = -2\cos x \sin x$

6.6 $\frac{dy}{dx} = \cos^2 x - \sin^2 x$

6.7 $\frac{dy}{dx} = 16\sin 8x \cos 8x$

6.8 $\frac{dy}{dx} = -15\cos^2(5x+1)\sin(5x+1)$

6.9 $\frac{dy}{dx} = \frac{-\sin x}{2\sqrt{1+\cos x}}$

6.10 $\frac{dy}{dx} = 2(\sin x + \cos x)(\cos x - \sin x)$

6.11 $\frac{dy}{dx} = x\sec x \tan x + \sec x$

6.12 $\frac{dy}{dx} = 4\sec^2 4x$

6.13 $\frac{dy}{dx} = \sec^3 x + \sec x \tan^2 x$

6.14 $\frac{dy}{dx} = 2\sec^2 x \tan x$

6.15 $\frac{dy}{dx} = 2\tan x \sec^2 x$

6.16 $\frac{dy}{dx} = 3\sec^3 x \tan x$

6.17 $\frac{dy}{dx} = \frac{(1+\cos x)\cos x + \sin^2 x}{(1+\cos x)^2} = \frac{1}{1+\cos x}$

6.18 $\frac{dy}{dx} = 2(\tan x + 1)\sec^2 x$

6.19 $\frac{dy}{dx} = -3(2x^2+1)\sin 3x + 4x\cos 3x$

6.20 $\frac{dy}{dx} = \frac{(5x\cos 5x - \sin 5x)}{x^2}$

6.21 $\frac{dy}{dx} = \frac{2x\sin x - (x^2+1)\cos x}{\sin^2 x}$

6.22 $\frac{dy}{dx} = \frac{\cos x(x\cos x + \sin x) + x\sin^2 x}{\cos^2 x}$
$= \frac{x + \sin x \cos x}{\cos^2 x}$

6.25 $\tan(x+y) = \frac{\sin(x+y)}{\cos(x+y)} = \frac{\sin x \cos y + \cos x \sin y}{\cos x \cos y - \sin x \sin y}$
Divide top and bottom by $\cos x \cos y$.

6.26 600 ft/min or 10 ft/sec

6.27 (i) $\omega = \frac{\pi}{2}$; $t = 3$ sec
(ii) $\frac{dx}{dt} = 0$ when $t = 1$, $\frac{dx}{dt} = -30$ in./sec at $t = 2$; $\frac{dx}{dt} = 0$ at $t = 3$

6.28 $\frac{d\theta}{dt} = \frac{3}{5}$ rad/sec

6.29 At $t = 3$, $\frac{d\theta}{dt} = \frac{44 \cdot 200}{(200^2 + 132^2)}$ rad/sec

6.33 $\frac{\pi}{2}, \frac{3\pi}{2}$

6.34 $0, \pi, 2\pi$

6.35 $\frac{\pi}{4}, \frac{5\pi}{4}$

6.36 $0, \frac{\pi}{2}, \pi, \frac{3\pi}{2}, 2\pi$

6.37 $0, \pi, 2\pi$

CHAPTER 7

7.1 1.16, 1.105, 1.099, 1.099

7.2 1.05, 1.005, 1.0004, 1

7.3 $\frac{dy}{dx} = 2e^{2x}$

7.4 $\frac{dy}{dx} = e^{x^2+x}(2x+1)$

7.5 $\frac{dy}{dx} = e^{-x^2}(-2x)$

7.6 $\frac{dy}{dx} = 2\left(e^{\sqrt{x}} + 1\right)\left(\frac{1}{2\sqrt{x}}e^{\sqrt{x}}\right)$

7.7 $\frac{dy}{dx} = \frac{1}{2}x^{\frac{1}{2}}e^{\frac{x}{2}} + \frac{1}{2}x^{-\frac{1}{2}}e^{\frac{x}{2}}$

7.8 $\frac{dy}{dx} = 3^x \log 3$

7.9 $\frac{dy}{dx} = 2^{x^2} \cdot 2x \log 2$

7.10 $\frac{dy}{dx} = 2xe^{3x} + 3x^2e^{3x} + \cos x - x \sin x$

7.11 $\frac{dy}{dx} = e^x \cos x - e^x \sin x - e^{-x} \sin x + e^{-x} \cos x$

7.12 $\frac{dy}{dx} = e^x \log x + \frac{1}{x}e^x$

7.13 $\frac{dy}{dx} = 2xe^{x^2} \log(1 + x^3) + \frac{e^{x^2} \cdot 3x^2}{1+x^3}$

7.14 $\frac{dy}{dx} = \frac{1}{x}$

7.15 $\frac{dy}{dx} = \frac{2x+1}{x^2+x}$

7.16 $\frac{dy}{dx} = \frac{1}{x} + 1$

7.17 $\frac{dy}{dx} = \frac{1}{x+1} + \frac{1}{x+2}$

7.18 $\frac{dy}{dx} = \frac{2}{x} - \log 3$

7.19 $\log y = x \log x$; $\frac{1}{y}\frac{dy}{dx} = \log x + 1$; $\frac{dy}{dx} = x^x(\log x + 1)$

7.20 $\frac{dy}{dx} = (2x+1)^x \left[\log(2x+1) + \frac{2x}{2x+1}\right]$

7.21 $\frac{dy}{dx} = (\sin x + \cos x)^x \left[\log(\sin x + \cos x) + \frac{x(\cos x - \sin x)}{\sin x + \cos x}\right]$

7.22 $A = 1000$; $k = \frac{1}{2}\log 3 = \log \sqrt{3}$; $y(4) = 9000$

7.24 $k = \log 2$

7.25 $\frac{\log 2}{\log(1.06)}$. Note $\log 2 \doteq .69$ and $\log(1 + x) \doteq x$ for small x so $\frac{\log 2}{\log(1+x)} \doteq \frac{.69}{x}$ (70 divided by the interest rate is the usual approximation).

7.26 $n = \log 2 / \log(1 + x) \doteq .70/x$

7.29 (b) $\left(\frac{1}{a}, e\right)$; (c) $x = \left(a^2e^2 + 1\right)/a$

7.30 $\left(-1, -\frac{1}{e}\right)$

CHAPTER 8

8.1 $\frac{dy}{dx} = \frac{2}{3}x^{-\frac{1}{3}}$

8.2 $\frac{dy}{dx} = -\frac{3}{4}x^{-\frac{7}{4}} + \frac{5}{4}x^{-\frac{3}{4}}$

8.3 $\frac{dy}{dx} = \frac{21}{5}x^{\frac{2}{5}} + \frac{14}{5}x^{-\frac{12}{5}}$

8.4 $\frac{dy}{dx} = \frac{3}{2}x^{\frac{1}{2}} - \frac{15}{2}x^{\frac{3}{2}}$

8.5 $\frac{dy}{dx} = -\frac{2}{3}(x^2 + 1)^{-\frac{5}{3}}(2x)$

8.6 $\frac{dy}{dx} = \left[\cos(x^{\frac{2}{3}} + 1)\right]\frac{2}{3}x^{-\frac{1}{3}}$

8.7 $\frac{dy}{dx} = x(1 + x^2)^{-\frac{1}{2}}$

8.8 $\frac{dy}{dx} = \frac{1}{2}(\cos^2 x + 1)^{-\frac{1}{2}}(-2\cos x \sin x)$

8.9 $\frac{dy}{dx} = \sec^2(x^{\frac{1}{3}} + 1)(\frac{1}{3}x^{-\frac{2}{3}})$

8.10 $\frac{dy}{dx} = -\frac{1}{3}(\sec x + 1)^{-\frac{4}{3}}(\sec x \tan x)$

8.11 $\sin^{-1}\frac{1}{2} = \frac{\pi}{6}$

8.12 $\cos^{-1}\frac{\sqrt{2}}{2} = \frac{\pi}{4}$

8.13 $\tan^{-1}\frac{1}{\sqrt{3}} = \frac{\pi}{6}$

8.14 $\cos^{-1}\frac{\sqrt{3}}{2} = \frac{\pi}{6}$

8.15 $\sin^{-1}\left(-\frac{\sqrt{3}}{2}\right) = -\frac{\pi}{3}$

8.16 $\cos^{-1}\left(-\frac{1}{2}\right) = \frac{2\pi}{3}$

8.17 $\cos^{-1}(-1) = \pi$

8.18 $\sin^{-1}(1) = \frac{\pi}{2}$

8.19 $\sin^{-1}(\sin\pi) = 0$

8.20 $\cos^{-1}(\cos(-\frac{\pi}{3})) = \frac{\pi}{3}$

8.21 $\frac{dy}{dx} = \frac{2x}{\sqrt{1-x^4}}$

8.22 $\frac{dy}{dx} = \frac{3}{\sqrt{1-9x^2}}$

8.23 $\frac{dy}{dx} = \frac{2}{\sqrt{1-(2x-1)^2}}$

8.24 $\frac{dy}{dx} = \frac{-1}{\sqrt{1-x}} \cdot \frac{1}{2\sqrt{x}}$

8.25 $\frac{dy}{dx} = \frac{-e^x}{\sqrt{1-e^{2x}}}$

8.26 $\frac{dy}{dx} = \frac{-1}{\sqrt{1-(x+1)^2}}$

8.27 $\frac{dy}{dx} = \frac{3}{1+9x^2}$

8.28 $\frac{dy}{dx} = \frac{1}{1+(x+2)^2}$

8.29 $\frac{dy}{dx} = \frac{\frac{1}{x}}{1+(\log x)^2}$

8.30 $\frac{dy}{dx} = 1$

8.34 $y = \frac{1}{5}\tan^{-1}\left(\frac{x-2}{5}\right)$

8.35 (a) $x/\sqrt{1-x^2}$
(b) $1/\sqrt{1+x^2}$

(c) $1/\sqrt{1-x^2}$
(d) x

CHAPTER 9

9.1 $f'(x) = 3x^2 - 10x + 3; \quad f''(x) = 6x - 10$
9.2 $f'(x) = 7x^6 - 6x^2 - 12x^{-5}; \quad f''(x) = 42x^5 - 12x + 60x^{-6}$
9.3 $f'(x) = \frac{1}{2}x^{-\frac{1}{2}} + \frac{1}{3}x^{-\frac{4}{3}}; \quad f''(x) = -\frac{1}{4}x^{-\frac{3}{2}} - \frac{4}{9}x^{-\frac{7}{3}}$
9.4 $f'(x) = \frac{-2x}{(1+x^2)^2}; \; f''(x) = \frac{-2(1+x^2)^2+8x^2(1+x^2)}{(1+x^2)^4}$
$= \frac{2(3x^2-1)}{(1+x^2)^3}$
9.5 $\frac{dy}{dx} = 2xe^{x^2}; \quad \frac{d^2y}{dx^2} = 2e^{x^2} + 4x^2e^{x^2}$
9.6 $\frac{dy}{dx} = 2\cos 2x + 3\sin 3x; \quad \frac{d^2y}{dx^2} = -4\sin 2x + 9\cos 3x$
9.7 $\frac{dy}{dx} = \frac{1}{\sqrt{1-x^2}}; \quad \frac{d^2y}{dx^2} = x(1-x^2)^{-\frac{3}{2}}$
9.8 $\frac{dy}{dx} = \frac{1}{1+\frac{x^2}{4}} \cdot \frac{1}{2}; \quad \frac{d^2y}{dx^2} = -4x(4+x^2)^{-2}$
9.9 local max at $x = -1$; local min at $x = 1$
9.10 local max at $x = -2$; local min at $x = 0$
9.11 local max at $x = -\frac{1}{2}$; local min at $x = \frac{1}{2}$
9.12 local max at $x = \frac{3}{4}$
9.13 local min at $x = -1$
9.14 local max at $x = 1$; local min at $x = 3$
9.17 10 in. by 10 in.
9.18 128 cu in.
9.19 $\frac{h}{r} = 2$
9.20 $\left(\frac{B}{2}\right)^2$
9.21 $x = \frac{2}{3}R, \quad h = \frac{1}{3}H$
9.22 $x = \frac{12}{13}, \quad y = \frac{21}{13}$
9.23 $x = 1, \quad y = 0$
9.24 $\ell = \left(h^{\frac{2}{3}} + b^{\frac{2}{3}}\right)^{\frac{3}{2}}$
9.25 $v = 7.5$ mph
9.26 A is minimum for $x = 100\pi/(4+\pi) \doteq 44$ in.
A is maximum for $x = 100$ in.

CHAPTER 10

10.1 $\sin\left(\frac{\pi}{6} - \frac{\pi}{180}\right) \doteq .5 - \frac{\sqrt{3}}{2} \cdot \frac{\pi}{180} = .4849$
10.2 $\sqrt{4.02} \doteq 2 + \frac{1}{2\sqrt{4}} \cdot (.02) = 2.005$
10.3 $e^{0.2} \doteq e^0 + e^0 \cdot (0.2) = 1.2$
10.4 $\tan^{-1}(1.04) \doteq \tan^{-1} 1 + \frac{1}{1+1^2} \cdot (.04) = .8054$
10.5 $\log(1.002) \doteq \log 1 + \frac{1}{1} \cdot (.002) = .002$
10.6 $2.01^3 \doteq 2^3 + 3 \cdot 2^2 \cdot (.01) = 8.12$
10.7 $30^{\frac{1}{5}} \doteq 32^{\frac{1}{5}} + \frac{1}{5}(32)^{\frac{-4}{5}} \cdot (-2) = 2 - \frac{2}{5 \cdot 16} = 1.975$
10.8 $f(3.1) \doteq f(3) + f'(3) \cdot (.1) = 25 + 2(.1) = 25.2$

10.9 $\frac{dW}{W} = 3\frac{dr}{r}$

10.10 $\frac{1}{3}$

10.11 $\frac{1}{2}$

10.12 $\frac{1}{2}$

10.13 1

10.14 $-\frac{1}{\pi}$

10.15 $\frac{1}{2}$

10.16 0

10.17 0

10.18 $x_2 = 2.25; \quad x_3 = 2.236$

10.19 $x_2 = 1.444; \quad x_3 = 1.4423$

10.20 .6823

10.21 .567

CHAPTER 11

11.1 $\frac{1}{4}x^4$

11.2 $-\frac{1}{3}x^{-3}$

11.3 $\frac{3}{10}x^{\frac{10}{3}}$

11.4 $2x^{\frac{1}{2}}$

11.5 $-10x^{-\frac{1}{2}}$

11.6 $4x^{\frac{7}{4}}$

11.7 $\frac{4}{3}x^{\frac{3}{2}} + \frac{3}{4}x^4$

11.8 $\frac{1}{2}x^2 + \frac{5}{4}x^4$

11.9 $-\frac{1}{3}\cos 3x$

11.10 $\frac{1}{5}\sin 5x$

11.11 $-2\cos\frac{x}{2} - \frac{3}{2}\sin 2x$

11.12 $\frac{1}{3}\sin x^3$

11.13 $-\cos(1+x)$

11.14 $\frac{1}{2}\sin(1+x^2)$

11.15 $\frac{2}{9}(4+x^3)^{\frac{3}{2}}$

11.16 $\frac{1}{22}(1+x^2)^{11}$

11.17 $\frac{3}{2}x^2 + \frac{5}{4}e^{4x}$

11.18 $x + 2e^x + \frac{1}{2}e^{2x}$

11.19 $\frac{1}{2}e^{x^2}$

11.20 $\sin e^x$

11.21 $x + 3\log|x|$

11.22 $\frac{2}{3}\log|3+x^3|$

11.23 $\log|x^2+x+5|$

11.24 $\frac{1}{2}\sin^{-1}(2x)$

11.25 $\sin^{-1}(\frac{x}{2})$

11.26 $\frac{5}{3}\tan^{-1}(3x)$
11.27 $\frac{1}{3}\tan^{-1}\frac{x}{3}$
11.28 $\log|x+5|$
11.29 $y = x^3 + \frac{1}{2}x^2 - 7x + \frac{15}{2}$
11.30 $y = e^x + \log|1+x| + 1$
11.31 $y = \tan^{-1}x - \frac{\pi}{4}$
11.32 $y = \sin^{-1}x + 5$
11.33 $k = 22$ ft/sec^2

CHAPTER 12

12.1 $\frac{38}{3}$
12.2 2
12.3 $\frac{152}{3}$
12.4 $\frac{3}{5}[3^{\frac{5}{3}} - 2^{\frac{5}{3}}] + \log\frac{2}{3}$
12.5 $\frac{56}{3}$
12.6 $2 - 2\sin\frac{1}{2}$
12.7 $\frac{1}{3}$
12.8 1
12.9 $\frac{1}{2}(e-1)$
12.10 $\frac{\pi}{4}$
12.11 $\frac{\pi}{6}$
12.12 $\frac{1}{2}\log 2$
12.13 36
12.14 2
12.15 $\frac{9}{2}$
12.16 $\frac{8}{3}$
12.17 $\frac{4}{3}$
12.18 $\frac{1}{4}$
12.19 $1 - \frac{\pi}{4}$
12.20 $\frac{17}{3}$
12.22 (i) e^{x^2}; (ii) $\sqrt{1+x^3}$; (iii) $\sin x^2$; (iv) $-\log(1+x^2)$
12.23 $f(g(x))g'(x)$

CHAPTER 13

13.1 12 ft lbs
13.2 35 ft lbs
13.3 2250 ft lbs
13.4 $\frac{1}{2}\cdot 62\cdot\pi\cdot 16\cdot 36$ ft lbs
13.5 2852 ft lbs
13.6 18 cubic units

13.7 $\frac{1}{3}\pi a^3$ cubic units

13.8 $\frac{4}{3}\pi a^3$ cubic units

13.9 $\frac{\pi}{5}$ cubic units

13.10 $\frac{\pi}{2}$ cubic units

13.11 $\frac{\pi^2}{4}$ cubic units

13.12 $\frac{4}{3}\pi ab^2$; $\frac{4}{3}\pi ba^2$ cubic units

13.14 264 ft $= \frac{1}{20}$ mi

13.15 9920 lbs

13.16 225 ft lbs

13.17 25 ft lbs

CHAPTER 14

14.1 The curve is the arc of the parabola $y = x^2$ between $(0, 0)$ and $(1, 1)$.

14.2 The curve is the arc of the parabola $(y + 1)^2 = 4x$ between $(0, -1)$ and $(1, 1)$.

14.3 The curve is the top half of the ellipse $\left(\frac{x}{2}\right)^2 + y^2 = 1$.

14.4 The curve is the arc of the hyperbola $y = \frac{1}{x}$ from $\left(\frac{1}{2}, 2\right)$ to $(1, 1)$.

14.5 $x = -y^2 + 2$; $1 \le x \le 2$

14.6 $y = 2x^2 - 1$; $x \ge 0$

14.7 $y = (x - 1)^2$; $-\infty < x < 1$

14.8 $y = 1 - x$; $0 \le x \le 1$

14.9 $y - 4 = \frac{1}{6}(x - 15)$; curve lies under the tangent line

14.10 $y = 1$; curve lies over the tangent line

14.11 $y = x + 2$; curve is the tangent line

14.12 $y - 1 = 2(x - 1)$; curve lies over tangent line

14.13 $\frac{dy}{dx} = 0$ when $t = \sqrt{2}$ sec and $y = 32$ ft

14.14 $t = v \sin\theta/32$ for max. height. $t = v \sin\theta/16$ when body hits the ground. When the curve hits the ground, $\frac{dy}{dx} = -\tan\theta$, so the angle of impact is $\pi - \theta$.

14.15 $2\pi a$

14.16 $\frac{728}{24} = \frac{91}{3}$

14.17 5

14.18 $\log\left|\sqrt{2} + 1\right|$

14.19 $\sqrt{2}\left(e^{\pi} - 1\right)$

14.20 $\sinh x_0$

14.21 $x = a\sec\theta$, $y = b\tan\theta$

14.22 $x = am$, $y = am^2$

14.24 $x = a\cos\theta$, $y = b\sin\theta$

14.25 $x = a\cos\theta$, $y = b\sin\theta$

CHAPTER 15

15.1 $\frac{1}{12}(4^6 - 3^6)$

15.2 $\frac{1}{2}\log\frac{7}{5}$

15.3 $\frac{26}{3}$
15.4 0
15.5 $\frac{1}{4}$
15.6 $\frac{2}{3}$
15.7 $\frac{3}{2}(\log 2)^2$
15.8 $\frac{15}{4}$
15.9 $4 + 2\log 5$
15.10 $\frac{3}{2}(e - 1)$
15.11 $2(e^2 - e)$
15.12 $\frac{\pi}{8}$
15.13 $\frac{1}{2}\left(\tan^{-1} 2\right)$
15.14 $\frac{1}{2}\log 10$
15.15 $\tan^{-1} e - \frac{\pi}{4}$
15.16 $\sin^{-1}\frac{1}{4}$
15.17 $\frac{2}{3}\left(\sqrt{5} - \sqrt{2}\right)$
15.18 0
15.19 $\frac{1}{6}\left(\tan^{-1}\frac{4}{3} - \tan^{-1}\frac{2}{3}\right)$
15.20 $\frac{1}{3}\sin^{-1}\frac{3}{4}$
15.21 $\frac{1}{10}\left(\tan^{-1}\frac{15}{2} - \tan^{-1}\frac{5}{2}\right)$
15.22 $\log 61 - \log 5$
15.23 $\frac{1}{12} + \frac{1}{11}$
15.24 $\frac{2}{5}\left(3^5 - 2^5\right) - \frac{4}{3}\left(3^3 - 2^3\right)$
15.25 $\frac{2}{3}\left(2^{\frac{3}{2}} - 1\right) + 2\left(2^{\frac{1}{2}} - 1\right)$
15.26 $\frac{3}{10} + \frac{6}{7} + \frac{3}{4}$
15.27 $\frac{18}{5}\sqrt{3} - \frac{64}{15}\sqrt{2}$

CHAPTER 16

16.1 $3\log|x + 2|$
16.2 $-\frac{1}{2}\frac{1}{(x-3)^2}$
16.3 $\frac{3}{4}\log|4x + 1|$
16.4 $x + 2\log|x - 1|$
16.5 $\frac{1}{2}x^2 + x + \log|x|$
16.6 $\frac{1}{2}x^2 - 4x + 16\log|x + 4|$
16.7 $\frac{2}{3}\log\left|\frac{x-5}{x-2}\right|$
16.8 $\frac{1}{3}\log\left|\frac{x-2}{x+1}\right|$
16.9 $\log\left|\frac{(x-2)^2}{x-1}\right|$
16.10 $\log\left|\frac{x-1}{x+1}\right|$
16.11 $\frac{1}{4}\log\left|\frac{x+2}{x-2}\right|$
16.12 $\tan^{-1}(x + 1)$
16.13 $\tan^{-1}(x + 2)$

16.14 $\frac{3}{2}\log\left|x^2+1\right|$
16.15 $\log\left|x^2+9\right|+\frac{1}{3}\tan^{-1}\frac{x}{3}$
16.16 $\frac{1}{2}\log\left|x^2+4\right|+\frac{1}{2}\tan^{-1}\frac{x}{2}$
16.17 $2\log\left|x^2+6x+10\right|-12\tan^{-1}(x+3)$
16.18 $\log\left|\frac{x+1}{x+2}\right|$
16.19 $\frac{3}{2}\log\left|x^2+4x+8\right|-6\tan^{-1}\frac{(x+2)}{2}$
16.20 $3\log|x+1|+\log\left|x^2+1\right|$
16.21 $4\log|x|+\log\left|x^2+2x+2\right|+\tan^{-1}(x+1)$
16.22 $\log\left|\frac{(x-1)(x+2)^2}{x-3}\right|$
16.23 $\log\left|\frac{x^2(x+1)}{x+2}\right|$
16.24 $\log|x|-\frac{1}{2}\log\left|x^2+4\right|$

CHAPTER 17

17.1 $x\cos^{-1}x+\sqrt{1-x^2}$
17.2 $x\tan^{-1}x-\frac{1}{2}\log|1+x^2|$
17.3 $-\frac{1}{2}xe^{-2x}-\frac{1}{4}e^{-2x}$
17.4 $x^2e^x-2xe^x+2e^x$
17.5 $-\frac{1}{2}x^3e^{-2x}-\frac{3}{4}x^2e^{-2x}-\frac{3}{4}xe^{-2x}-\frac{3}{8}e^{-2x}$
17.6 $-x\cos x+\sin x$
17.7 $\frac{1}{3}x\sin 3x+\frac{1}{9}\cos 3x$
17.8 $-\frac{1}{5}x^2\cos 5x+\frac{2}{25}x\sin 5x+\frac{2}{125}\cos 5x$
17.9 $\frac{1}{2}x^2\log x-\frac{1}{4}x^2$
17.10 $\frac{2}{3}x^{\frac{3}{2}}\log x-\frac{4}{9}x^{\frac{3}{2}}$
17.11 $\frac{1}{8}x^8\log x-\frac{1}{64}x^8$
17.12 $x(\log x)^2-2x\log x+2x$
17.13 $x(\log x)^3-3x(\log x)^2+6x\log x-6x$
17.14 $\frac{2}{3}x(1+x)^{\frac{3}{2}}-\frac{4}{15}(1+x)^{\frac{5}{2}}$
17.15 $x\log(1+x^2)-2x+2\tan^{-1}x$
17.16 $\frac{1}{2}e^x(\sin x-\cos x)$
17.17 $2\sqrt{x+1}\left[\log(x+1)-2\right]$
17.18 $-\frac{1}{2}e^{-x^2}\left(x^2+1\right)$
17.19 $\frac{\pi}{2}-1$
17.20 $\frac{1}{4}$
17.21 $\frac{1}{15}\left[32+16\sqrt{2}\right]$
17.22 $\frac{\pi}{24}+\frac{\sqrt{3}}{4}-\frac{1}{2}$
17.23 $\frac{\pi}{2}\left(e^2+1\right)$
17.24 2π
17.25 $\frac{2}{3}\pi a^3$

CHAPTER 18

18.1 $\frac{1}{4}\sin^4 x$

18.2 $\frac{2}{5}\sin^{\frac{5}{2}} x$

18.3 $-\frac{1}{6}\cos^6 x$

18.4 $\log|\sin x|$

18.5 $\sin x - \frac{1}{3}\sin^3 x$

18.6 $-\frac{1}{4}\cos^4 x$

18.7 $-\cos x + \frac{1}{3}\cos^3 x$

18.8 $-\frac{2}{3}\cos^{\frac{3}{2}} x$

18.9 $\frac{1}{3}\cos^3 x - \frac{1}{5}\cos^5 x$

18.10 $\frac{1}{2}x - \frac{1}{2}\sin x\cos x$

18.11 $\frac{1}{8}x - \frac{1}{8}\sin x\cos^3 x + \frac{1}{8}\sin^3 x\cos x$

18.12 $\frac{3}{8}x + \frac{1}{4}\sin 2x + \frac{1}{32}\sin 4x$

18.13 $\frac{1}{16}x - \frac{1}{64}\sin 2x + \frac{1}{48}\sin^3 2x$

18.14 $\frac{1}{2}\tan 2x$

18.15 $\frac{1}{3}\tan^3 x$

18.16 $\tan x - x$

18.17 $\frac{1}{12}\sec^4 3x$

18.18 $-\cos x$

18.19 $\frac{2}{3}\tan^{\frac{3}{2}} x$

18.20 $\frac{1}{2}\tan^2 x + \log|\cos x|$

18.21 $\log|\sec x + \tan x| - \log|\cos x|$

18.22 $\frac{1}{8}[\sec 4x\tan 4x + \log|\sec 4x + \tan 4x|]$

18.23 $\frac{1}{4}[\sec 2x\tan 2x + \log|\sec 2x + \tan 2x|]$

18.24 $\sec x$

18.26 (ii) $-\frac{1}{4}\sin^3 x\cos x - \frac{3}{8}\sin x\cos x + \frac{3}{8}x$

18.27 $\frac{\pi}{16}$

18.28 $\frac{3}{8}\pi^2$

18.29 $2\sqrt{2}$

18.32 (i) $-\frac{1}{10}\cos 5x - \frac{1}{2}\cos x$
(ii) $-\frac{1}{12}\cos 6x + \frac{1}{8}\cos 4x$

CHAPTER 19

19.1 $\frac{\pi}{2}$

19.2 $\frac{1}{2}\sin^{-1} x + \frac{1}{2}x\sqrt{1-x^2}$

19.3 $2\sin^{-1}\frac{x}{2} + \frac{1}{2}x\sqrt{4-x^2}$

19.4 $\frac{1}{3}$

19.5 $-\sqrt{1-x^2}$

19.6 $\frac{x}{\sqrt{1-x^2}} - \sin^{-1} x$

19.7 $\frac{1}{9}\frac{x}{\sqrt{9-x^2}}$

19.8 $\sin^{-1}\frac{x}{\sqrt{3}}$

19.9 $-\frac{1}{a}\log\left|\frac{a+\sqrt{a^2-x^2}}{x}\right|$

19.10 πab

19.11 $\log\left|\sqrt{1+x^2}+x\right|$

19.12 $\sqrt{4+x^2}$

19.13 $-\frac{\sqrt{1+x^2}}{x}$

19.14 $\sqrt{x^2-4} - 2\sec^{-1}\frac{x}{2}$

19.15 $\frac{1}{2}\sec^{-1}\frac{x}{2}$

19.16 $\frac{1}{2}x\sqrt{x^2-9} + \frac{9}{2}\log\left|x+\sqrt{x^2-9}\right|$

19.17 $\frac{1}{2}x\sqrt{x^2-1} - \frac{1}{2}\log\left|x+\sqrt{x^2-1}\right|$

19.18 $\frac{1}{2}\left[\tan^{-1}x + \frac{x}{1+x^2}\right]$

19.19 $-\frac{1}{2}\frac{1}{1+x^2} + \frac{1}{2}\left[\tan^{-1}x + \frac{x}{1+x^2}\right]$

19.21 $y = a\log\left|\frac{a-\sqrt{a^2-x^2}}{x}\right| - \sqrt{a^2-x^2}$

19.26 $\cosh^{-1}u = \log\left|u+\sqrt{u^2-1}\right|$

CHAPTER 20

20.1 $S_4 = .69325$; $S_{10} = .693150$

20.2 $S_{10} = .7853982$

20.3 $S_{10} = .7818$

20.4 (i) $\frac{9}{2}\left[\frac{\sqrt{10}}{9} + \log\left(\frac{\sqrt{10}+1}{3}\right)\right]$
(ii) 3.0546645
(iii) $S_{10} = 3.0546645$

20.5 $S_4 = .8285$; $S_8 = .8282$

CHAPTER 21

21.1 −1

21.2 0

21.3 0

21.4 0

21.5 0

21.6 0

21.7 0

21.8 0

21.9 0

21.10 0

21.11 1

21.12 1

21.13 0

21.14 ∞

21.15 0

21.16 e^2

21.17 0

21.18 ∞

21.19 0

21.20 ∞

21.21 $\frac{1}{n}|x|^n \longrightarrow 0$ if $|x| \leq 1$; $\frac{1}{n}|x|^n \longrightarrow \infty$ if $|x| > 1$

21.22 $n^3|x|^n \longrightarrow 0$ if $|x| < 1$; $n^3|x|^n \longrightarrow \infty$ if $|x| \geq 1$

21.23 $\frac{(\log n)|x|^n}{n^2} \longrightarrow 0$ if $|x| \leq 1$; $\frac{(\log n)|x|^n}{n^2} \longrightarrow \infty$ if $|x| > 1$

21.24 $\frac{2^n|x|^n}{n^3} \longrightarrow 0$ if $|x| \leq \frac{1}{2}$; $\frac{2^n|x|^n}{n^3} \longrightarrow \infty$ if $|x| > \frac{1}{2}$

21.25 $\frac{(n^2+1)|x|^n}{3^n+n} \longrightarrow 0$ if $|x| < 3$; $\frac{(n^2+1)|x|^n}{3^n+n} \longrightarrow \infty$ if $|x| \geq 3$

21.26 $n!|x|^n \longrightarrow 0$ if $x = 0$; $n!|x|^n \longrightarrow \infty$ if $x \neq 0$

21.27 $\frac{n!|x|^n}{(n+2)!} \longrightarrow 0$ if $|x| \leq 1$; $\frac{n!|x|^n}{(n+2)!} \longrightarrow \infty$ if $|x| > 1$

21.28 $\frac{(n+2)!|x|^n}{n!} \longrightarrow 0$ if $|x| < 1$; $\frac{(n+2)!|x|^n}{n!} \longrightarrow \infty$ if $|x| \geq 1$

CHAPTER 22

22.1 1

22.2 $\frac{3}{8}$

22.3 $\frac{\pi}{2}$

22.4 diverges

22.5 1

22.6 $\frac{1}{2}$

22.7 diverges

22.8 diverges

22.9 diverges

22.10 2

22.11 diverges

22.12 $\frac{\pi}{2}$

22.13 $\frac{2}{e}$

22.14 $\frac{\pi}{12}$

22.15 diverges

22.16 $\frac{-2}{1+\sqrt{2}}$

22.17 converges

22.18 converges

22.19 converges

22.20 converges

22.21 π

22.22 $\frac{\pi}{2e^2}$

22.23 $\frac{\pi^2}{2}$
22.24 $\frac{\pi}{4}$
22.25 (iii) $\int_0^\infty x^2 e^{-x}\,dx = 2, \quad \int_0^\infty x^3 e^{-x}\,dx = 3 \cdot 2$
$\int_0^\infty x^4 e^{-x}\,dx = 4 \cdot 3 \cdot 2, \cdots, \int_0^\infty x^n e^{-x}\,dx = n!$

CHAPTER 23

23.1 500
23.2 $\frac{1}{2} \cdot 100 \cdot 101 = 5050$
23.3 210
23.4 100
23.5 N^2
23.6 $1 - \frac{1}{3^{10}}$
23.7 $72.38672 = \frac{\left[\left(\frac{3}{2}\right)^2 - \left(\frac{3}{2}\right)^9\right]}{\left(1-\frac{3}{2}\right)}$
23.8 diverges; $\frac{n}{n+1} \longrightarrow 1$
23.9 converges; proper alternating series
23.10 converges; proper alternating series
23.11 converges; proper alternating series
23.12 diverges; $\frac{1}{\log n} > \frac{1}{n}$
23.13 diverges; $\frac{\log n}{n} > \frac{1}{n}$
23.14 converges; geometric series
23.15 converges; integral test
23.16 diverges; $\frac{e^n}{n^2} \longrightarrow \infty$
23.17 diverges; integral test
23.18 diverges; integral test
23.19 converges; $\frac{1}{n} \sin \frac{1}{n} < \frac{1}{n^2}$
23.20 diverges; $\frac{4^n}{n 3^n} \longrightarrow \infty$
23.21 diverges; $e^{\frac{1}{n}} \longrightarrow 1$
23.22 converges; $\frac{2^n}{3^n + 4^n} < \left(\frac{2}{3}\right)^n$
23.23 $a^2 < a$ if $0 < a < 1$
23.24 $\sin a < a$ if $0 < a < 1$

CHAPTER 24

24.1 1
24.2 $\frac{1}{2}$
24.3 4
24.4 $\frac{1}{2}$
24.5 5
24.6 1
24.7 ∞
24.8 ∞

24.9 ∞
24.10 $\frac{1}{e}$
24.11 $\frac{2}{3}$
24.12 1
24.13 10
24.14 1
24.15 $(2, 4)$
24.16 $\left(\frac{1}{2}, \frac{3}{2}\right)$
24.17 $(-\infty, \infty)$
24.18 $(-6, 4)$
24.19 $e^{2x} = 1 + 2x + \frac{1}{2!}(2x)^2 + \frac{1}{3!}(2x)^3 + \cdots; \quad (-\infty, \infty)$
24.20 $e^{-x} = 1 - x + \frac{1}{2!}x^2 - \frac{1}{3!}x^3 + \cdots; \quad (-\infty, \infty)$
24.21 $\log(1 - x) = -x - \frac{1}{2}x^2 - \frac{1}{3}x^3 - \frac{1}{4}x^4 - \cdots; \quad (-1, 1)$
24.22 $\log(1 + 2x) = 2x - 2x^2 + \frac{2^3}{3}x^3 - \frac{2^4}{4}x^4 + \frac{2^5}{5}x^5 - \cdots; \quad (-\frac{1}{2}, \frac{1}{2})$
24.23 $\frac{1}{(1+x)^2} = 1 - 2x + 3x^2 - 4x^3 + \cdots; \quad (-1, 1)$
24.24 $\frac{1}{1+x^3} = 1 - x^3 + x^6 - x^9 + x^{12} - \cdots; \quad (-1, 1)$
24.25 $\sin 2x = 2x - \frac{1}{3!}(2x)^3 + \frac{1}{5!}(2x)^5 - \cdots; \quad (-\infty, \infty)$
24.26 $\tan^{-1}\frac{x}{2} = \frac{x}{2} - \frac{1}{3}(\frac{x}{2})^3 + \frac{1}{5}(\frac{x}{2})^5 - \frac{1}{7}(\frac{x}{2})^7 + \cdots; \quad (-2, 2)$
24.27 $\cos x^2 = 1 - \frac{1}{2!}x^4 + \frac{1}{4!}x^8 - \frac{1}{6!}x^{12} + \cdots; \quad (-\infty, \infty)$
24.28 $\frac{x}{(1+x)^2} = x - 2x^2 + 3x^3 - 4x^4 + \cdots; \quad (-1, 1)$
24.29 $\frac{\sin x}{x} = 1 - \frac{x^2}{3!} + \frac{x^4}{5!} - \frac{x^6}{7!} + \cdots; \quad (-\infty, \infty)$
24.30 $\frac{x}{1+x^2} = x - x^3 + x^5 - x^7 + \cdots; \quad (-1, 1)$
24.31 $\log(1 + x^2) = x^2 - \frac{1}{2}x^4 + \frac{1}{3}x^6 - \frac{1}{4}x^8 + \cdots; \quad (-1, 1)$

CHAPTER 25

25.1 $P_1(x) = -1 + 3x$; $P_2(x) = -1 + 3x - 2x^2$; $P_3(x) = -1 + 3x - 2x^2 + 5x^3$; $P_4(x) = P_3(x)$
25.2 $P_4(x) = 1 - x + x^2 - x^3 + x^4$
25.3 $P_3(x) = x - \frac{1}{3}x^3$
25.4 $P_4(x) = x^2$; answer to (b) is the same
25.5 $P_5(x) = 1 + \frac{1}{2}(x - 1) + \frac{1}{2!}\left(-\frac{1}{2^2}\right)(x - 1)^2 + \frac{1}{3!}\frac{3}{2^3}(x - 1)^3$
$+\frac{1}{4!}\frac{(-5\cdot 3)}{2^4}(x - 1)^4 + \frac{1}{5!}\frac{(7\cdot 5\cdot 3)}{2^5}(x - 1)^5$
25.6 $P_4\left(\frac{1}{3}\right) = 1.39558$
$R_4\left(\frac{1}{3}\right) \leq \frac{2}{5!}\left(\frac{1}{3}\right)^5 = .00007$
25.7 $\int_0^1 P_3(x)\,dx = \frac{13}{42} = .3095$
$R_4(x^2) \leq \frac{1}{5!}x^{10}$; $\int_0^1 R_4(x^2)\,dx \leq \frac{1}{11\cdot 5!} = .0008.$
25.8 $\int_0^1 P_4(\sqrt{x})\,dx = \int_0^1 P_5(\sqrt{x})\,dx = .7639$;
$\left|R_4(\sqrt{x})\right| = \left|R_5(\sqrt{x})\right| \leq \frac{1}{6!}|x|^3$;
$\int_0^1 \frac{1}{6!}x^3\,dx = .0003$;
$\int_0^1 \cos\sqrt{x}\,dx = .7639 \pm .0003.$

25.9 $S_{10} = 1.4627$

25.10 $\int_0^1 P_5(x^2)\,dx = 1.4625$

$1.4625 + .0001 < \int_0^1 e^{x^2}\,dx < 1.4625 + .0003$

CHAPTER 26

26.1 $\cos\frac{1}{4} \doteq .9688$. Two terms are enough.

26.2 $\sin\frac{1}{10} \doteq .0998333 \pm 10^{-7}$

26.3 $.7786 < e^{-\frac{1}{4}} < .7788$

26.4 $P_4(x) = 1 + \frac{1}{3}x - \frac{1}{2!}\cdot\frac{2}{3^2}x^2 + \frac{1}{3!}\frac{2\cdot 5}{3^3}x^3 - \frac{1}{4!}\frac{2\cdot 5\cdot 8}{3^4}x^4$

26.5 $(1-x)^{\frac{1}{2}} = 1 - \frac{1}{2}x - \frac{1}{2!}\frac{1}{2^2}x^2 - \frac{1}{3!}\frac{1\cdot 3}{2^3}x^3 - \frac{1}{4!}\frac{1\cdot 3\cdot 5}{2^4}x^4 - \cdots$

$(1+x^2)^{\frac{1}{2}} = 1 + \frac{1}{2}x^2 - \frac{1}{2!}\frac{1}{2^2}x^4 + \frac{1}{3!}\frac{1\cdot 3}{2^3}x^6 - \frac{1}{4!}\frac{1\cdot 3\cdot 5}{2^4}x^8 + \cdots$

26.6 $e^x = e^3 + e^3(x-3) + \frac{e^3}{2!}(x-3)^2 + \frac{e^3}{3!}(x-3)^3 + \cdots$

$e^x = e^{-1} + e^{-1}(x+1) + \frac{e^{-1}}{2!}(x+1)^2 + \frac{e^{-1}}{3!}(x+1)^3 + \cdots$

26.7 $\frac{1}{x} = \frac{1}{2} - \frac{1}{2^2}(x-2) + \frac{1}{2^3}(x-2)^2 - \frac{1}{2^4}(x-2)^3 + \cdots$

26.8 $\log x = \log 2 + \frac{1}{2}(x-2) - \frac{1}{2}\cdot\frac{1}{2^2}(x-2)^2 + \frac{1}{3}\cdot\frac{1}{2^3}(x-2)^3 - \frac{1}{4}\cdot\frac{1}{2^4}(x-2)^4 + \cdots$

26.9 $\frac{\cos x - 1}{x^2} = -\frac{1}{2!} + \frac{x^2}{4!} - \frac{x^4}{6!} + \frac{x^6}{8!} - \cdots$

26.10 $.94608 < \int_0^1 \frac{\sin x}{x}\,dx < .94611$

CHAPTER 27

27.1 $y = \frac{1}{2}\sin x^2 + 2$

27.2 $y = \tan x - 1$

27.3 $y = \tan^{-1}\frac{x}{3} + \frac{\pi}{4}$

27.4 $y = \frac{1}{2}\log\left(x^2+1\right) + 5$

27.5 $\frac{1}{2}y^2 = \frac{1}{3}x^3 + x + 2$

27.6 $y = \log\left(e^x + 1\right)$

27.7 $\frac{1}{2}\tan^{-1}\frac{y}{2} = \tan^{-1}x + c$

27.8 $\log|y| = \sin^{-1}x + c$

27.9 $y = ce^{5x}$

27.10 $y = -2 + ce^{5x}$

27.11 $y = ce^{x^2}$

27.12 $y = -1 + ce^{x^2}$

27.13 $y = cxe^x$

27.14 $y = -2 + c(x-1)$

27.15 $T = 20 + 80\left(\frac{1}{4}\right)^{\frac{t}{5}}$; $T(10) = 25$

27.16 $y = 100(5)^{\frac{t}{3}}$; $y = 1000$ at $t = 4.29$ hrs; $y = 2500$ at $t = 6$ hrs

CHAPTER 28

28.1 $y = Cx^{-1}$

28.2 $y = \frac{1}{3}x^2 + Cx^{-1}$

28.3 $y = Ce^{-x^2}$

28.4 $y = \frac{1}{2} + Ce^{-x^2}$
28.5 $y = \frac{1}{2}e^x + Ce^{-x}$
28.6 $y = 2xe^{-x} + Ce^{-x}$
28.7 $y = -2x - 1 + Ce^{3x}$
28.8 $y = 2x^2 + Ce^{-2x}$
28.9 $y = \cos 3x + 2\sin 3x + Ce^{6x}$
28.10 $y = \frac{1}{5}\cos 2x + \frac{2}{5}\sin 2x + Ce^{-x}$
28.11 $y = \frac{1}{2}e^x + \frac{1}{5}\cos 2x + \frac{2}{5}\sin 2x + Ce^{-x}$
28.12 $y = -x - \frac{1}{6} + \cos 3x - 2\sin 3x + Ce^{6x}$
28.13 $y = e^{2x} + Ce^{-3x}$
28.15 $v = 13.8$ ft/sec
28.16 $T = 50.6$

CHAPTER 29

29.2 $y = C_1e^{-3t} + C_2e^{-2t}; \quad y = -7e^{-3t} + 8e^{-2t}$
29.3 $y = C_1e^{2t} + C_2e^{-t}; \quad y = 3e^{2t} - e^{-t}$
29.4 $y = C_1e^{-2t} + C_2e^{-t}; \quad y = -e^{-t}$
29.5 $y = C_1e^{3t} + C_2e^{-t}; \quad y = 2e^{3t} + e^{-t}$
29.6 $\frac{d^2y}{dt^2} - 2r\frac{dy}{dt} + r^2y = 0$
29.7 $y = C_1e^t + C_2te^t$
29.8 $y = C_1e^{-5t} + C_2te^{-5t}$
29.10 $y = C_1e^t\cos t + C_2e^t\sin t; \quad y = 3e^t\cos t + 7e^t\sin t$
29.11 $y = C_1e^t\cos 2t + C_2e^t\sin 2t; \quad y = 3e^t\cos 2t + 4e^t\sin 2t$
29.12 $y = C_1e^{3t}\cos t + C_2e^{3t}\sin t; \quad y = 6e^{3t}\sin t$
29.13 $y = C_1e^{2t}\cos 4t + C_2e^{2t}\sin 4t; \quad y = e^{2t}\cos 4t$
29.14 $y = C_1\cos 2t + C_2\sin 2t$; frequency $= \frac{1}{\pi}$
29.15 $y = 2\sin\left(2t + \frac{\pi}{3}\right)$
29.16 (a) $\ell = 3.24$ft; (b) $\theta = .158\sin\pi t$, $.158\text{rad} = 9.01°$

CHAPTER 30

30.1 (b) $\int te^{ct}dt = \frac{e^{ct}}{c^2}[ct - 1]$
$\int t^2e^{ct}dt = \frac{e^{ct}}{c^3}[c^2t^2 - 2ct + 2]$
$\int t^3e^{ct}dt = \frac{e^{ct}}{c^4}[c^3t^3 - 3c^2t^2 + 6ct - 6]$
30.2 $y = e^{-t}$
30.3 $y = -te^t$
30.4 $y = (2t^2 - 6t + 7)e^{3t}$
30.5 $y = (3t^2 + 2t)e^{2t}$
30.6 $y = 2t + 3$
30.7 $y = -3\sin t + \cos t + C_1e^{2t} + C_2e^{-t}$
30.8 $y = 3t\sin t + C_1\cos t + C_2\sin t$
30.9 $y = -4\cos 2t + 3\sin 2t + C_1e^t + C_2te^t$
30.10 $y = \frac{1}{2}te^t\sin t + C_1e^t\cos t + C_2e^t\sin t$

Index

About the Author

H. S. Bear received both his Bachelor of Arts degree and doctorate from the University of California at Berkeley. Prior to receiving his Ph.D., he taught for one year as an instructor at the University of Oregon. He then returned to the University of California at Berkeley and obtained his doctorate while teaching there. Dr. Bear has taught at a geographically diverse group of schools, including permanent positions at the University of Washington, University of California at Santa Barbara, and New Mexico State University. He has also held teaching positions at Princeton University, the University of California at San Diego, and the University of Erlangen-Nuremberg, Germany. Dr. Bear spent most of his academic career as professor of mathematics at the University of Hawaii, where he also served terms as chairman and graduate chairman of the Mathematics Department. He now holds the title of professor emeritus.

Throughout his career, Dr. Bear has been active in publishing books and articles in mathematics. His most recent publications are two advanced texts on mathematical analysis with Academic Press and a Dover Publications reprint of an earlier text on differential equations. With the publication of *Understanding Calculus: A User's Guide*, Dr. Bear will have contributed more than 40 pieces of mathematical literature.